Astrobiology and Planetary Sciences in Latin America

David Tovar • María Angélica Leal

Editors

Astrobiology and Planetary Sciences in Latin America

Research, Education, Public Outreach, and its Potential for Near-Future Investigations.

 Springer

Editors
David Tovar (iD)
Planetary Sciences and Astrobiology
Research Group
Universidad Nacional de Colombia /
Corporación Científica Laguna
Bogotá, BOGOTA, Colombia

María Angélica Leal
Planetary Sciences and Astrobiology
Research Group (GCPA)
Universidad Nacional de Colombia /
Corporación Científica Laguna
Bogotá, BOGOTA, Colombia

ISBN 978-3-032-01449-8   ISBN 978-3-032-01450-4   (eBook)
https://doi.org/10.1007/978-3-032-01450-4

Foreword

In recent decades, astrobiology has emerged as one of the most exciting and integrative fields in contemporary science. Arising at the intersection of biology, geology, astronomy, physics, and chemistry, this branch of knowledge dares to confront some of humanity's most profound questions: When, where, and how did life originate? Is life a phenomenon exclusive to our planet, or a possibility distributed throughout the universe? What conditions render a world habitable? How might we detect life beyond Earth?

Astrobiology is a profoundly multidisciplinary discipline that explores the origin, evolution, distribution, and potential destiny of life in the universe. Although there is still no unanimous definition of what exactly constitutes life, biology has identified its essential components—the building blocks of living systems—as well as the extraordinary variety of environments in which life can thrive. When the focus shifts to our immediate neighborhood—the planets and moons of the Solar System—astrobiology relies on planetary sciences. These disciplines, which take Earth as their primary reference point, enable the construction of atmospheric models, the performance of geodynamic simulations, the interpretation of imagery and chemical spectra, and the analysis of surface features, all with the aim of assessing whether a given environment could have once supported, or may still support, life on the surface or within the interior of a Solar System body. When astrobiology looks much farther afield—toward exoplanets or even regions of the interstellar medium—it turns to astrophysics and astrochemistry. These fields make it possible to investigate whether some of the thousands of planets discovered to date, or certain interstellar environments, might harbor forms of life similar to, or perhaps radically different from, those known on Earth.

What is perhaps most fascinating about astrobiology is not only the frontiers of knowledge it explores but also the profound effect it has on those who engage with it. It is a science that inspires, that awakens vocations, that opens the door to wonder—and leaves no one indifferent. Images of robotic vehicles traversing the Martian surface, spectacular photographs of distant bodies at the edge of our Solar System, the discovery of exoplanets in habitable zones, or the resilience of microbes in Earth's extreme environments are not merely technical and scientific achievements; they are also visual and symbolic narratives that nourish the collective imagination and strengthen the bond between science and society.

In Latin America, astrobiology is taking root with increasing strength. Dedicated research groups have emerged, academic programs have been launched, regional meetings are being organized, and a growing number of institutions across the region are participating in international networks and missions. What is most striking is that much of this development has not relied on large budgets or centralized infrastructures. Rather, it has been the result of the sustained efforts of small, inspired, and enthusiastic groups with a deep scientific, educational, and social commitment. Public universities, research centers, planetariums, museums, citizen associations, and dedicated educators have all contributed to driving this momentum forward.

But this is not all—the continent's geographical diversity offers an exceptional framework for studying life under extreme conditions and, by extension, for exploring its potential existence beyond Earth. The combination of inhospitable landscapes, unique ecosystems, and active geological processes makes this region a true natural laboratory, ideally suited for astrobiological and planetary research. Numerous sites closely replicate, with remarkable fidelity, environments that may be found on celestial bodies, such as Mars, Europa, or Enceladus.

In the Andes and the Southern Cone, numerous examples of such environments can be found. The Uyuni and Coipasa salt flats in Bolivia, rich in evaporites and exposed to intense ultraviolet radiation, have been studied as Martian analogs. In Chile, the Atacama Desert—the driest on Earth—serves as a privileged site for testing life-detection technologies and understanding the resilience of the most extreme ecosystems. Argentina adds to this landscape with arid northwestern regions, volcanic environments such as the Payunia field, and the hypersaline lagoons of the Puna, all of which offer conditions similar to those expected on Mars. In Brazil, the iron-rich outcrops of the Quadrilátero Ferrífero, together with the ecosystems of El Cerrado, have been used to investigate analogs from both a mineralogical and biological adaptation perspective in highly demanding environments. In Colombia, the Andean páramos and volcanoes such as Nevado del Ruíz provide opportunities to explore arid and hydrothermal settings and study endolithic organisms. Peru stands out for its retreating glaciers and high-altitude hypersaline lakes, which offer ideal conditions for investigating the cryospheres of other planetary bodies. Ecuador, with volcanoes such as Cotopaxi and Chimborazo, presents acidic and high-altitude conditions that facilitate the study of extreme geochemical processes. In Venezuela, the rock formations of the Guayana Shield and the Roraima tepuis—with their isolated ecosystems and unique biota—evoke truly extraterrestrial landscapes.

North and Central America also offer environments of high astrobiological value. In Mexico, the cenotes and caves of the Yucatán Peninsula, as well as the volcanoes of the Trans-Mexican Volcanic Belt, have served as key references for studying subsurface habitats and potential sheltered niches similar to those that may exist within lava tubes beneath the Martian surface. In Cuba, karst systems and low-light humid caves provide useful models for exploring radiation-protected environments. Costa Rica and Nicaragua, with active volcanoes such as Turrialba and Masaya, allow for the analysis of interactions between geothermal activity, gases, and microbial life. In Panama, the isthmus offers transitional zones between marine and terrestrial environments that are valuable for studying biosignatures in ecological frontier settings. In Guatemala, volcanoes such as Pacaya and Fuego provide dynamic contexts with significant atmospheric variability. Honduras and El Salvador have begun documenting craters and geological structures that are generating interest within the astrobiological community. Meanwhile, in the Dominican Republic, saline lakes and coastal ecosystems offer valuable conditions for studying biological adaptation and hydrological fluctuations—phenomena that may have occurred, for example, on Mars.

And one key factor cannot be overlooked: many Latin American countries are signatories to the Antarctic Treaty, which grants them access to a completely different and scientifically invaluable environment. In Antarctica, research is conducted in fields, such as microbiology in extreme environments, glacier dynamics, and permafrost processes—studies that enhance our understanding of the interactions between climate, life, ice, water, and rock. These investigations are of tremendous value for modeling potential ecosystems on icy moons, such as Europa or Enceladus, or even on frozen regions like the surface of Mars.

All of this environmental richness should not be viewed solely as a natural heritage. It is, above all, a scientific opportunity of the highest order. To seize it requires careful stewardship, responsible study, and the transformation of these assets into a strategic platform from which researchers in Latin America can actively contribute to planetary exploration and the advancement of astrobiology.

In this context, the purpose of this book is clear: to highlight, connect, and project the efforts currently underway in the region in the fields of astrobiology and planetary sciences. The aim is not merely to compile scattered experiences, but to construct a strategic and referential document that reflects the diverse trajectories, major accomplishments, and shared challenges—ultimately contributing to the consolidation of a cohesive regional scientific community with its own distinct identity.

One of the central elements of this volume is the recognition of pioneering groups that, across different countries and often with very limited resources, have contributed to key areas, such as planetary geology, microbiology in extreme environments, scientific instrumentation, and public outreach. This book also underscores the transformative role these disciplines have played in education, not only within formal school and university settings but also in less conventional environments, such as planetariums, museums, and citizen associations.

Furthermore, this book aims to serve as a practical tool for fostering both regional and international collaboration by providing concrete examples that may be useful to science policymakers, academic administrators, and institutions interested in strengthening their networks. Above all, this book is an invitation—a motivation for others to join the growing field of astrobiological research and to contribute their experience to future editions.

The first part of the volume provides a global overview of the development of astrobiology, featuring contributions from internationally recognized researchers. It addresses key topics, such as institutional consolidation, the role of international organizations like the IAU, and the impact of scientific events, networks, publications, and educational programs. The chapter dedicated to the Spanish Center for Astrobiology (CAB) illustrates how a single institution can successfully integrate research, training, participation in space missions, and international cooperation within a unified operational framework. Its experience offers a valuable model that may serve as inspiration for similar initiatives in Latin America.

The core of this book, however, is devoted to the national experiences of countries, such as Argentina, Brazil, Chile, Colombia, Cuba, Mexico, and Peru. These chapters showcase both well-established scientific trajectories and more recent initiatives. In all cases, particular emphasis is placed on the role that this discipline has played in higher education, science communication, and the cultivation of new scientific vocations—from amateur associations and school-based research projects to university programs focused on planetary sciences. These pages also document the emergence of collaborative networks, such as the Latin American Astrobiology Network, founded at the Second Latin American Congress in 2018, as well as national scientific societies that have helped to bring visibility to key actors, coordinate research agendas, and strengthen regional cooperation.

Although each country has followed its own path, the experiences gathered in this volume reveal common challenges across the region. The first is ensuring that public administrations recognize the strategic importance of astrobiology and planetary sciences, incorporating them into national science and technology agendas, allocating the necessary resources, and supporting the creation of networks, research centers, and programs that enable countries to position themselves internationally in the field of planetary exploration. The second major challenge is the consolidation of human capital. This involves improving both initial and continuing education for educators, creating joint postgraduate programs among institutions and countries, facilitating academic mobility, and ensuring stable working conditions with clear prospects for early-career researchers. The third challenge is strengthening scientific infrastructure—not only through the establishment of necessary laboratories, but also by providing the logistical means required for this kind of work. While the region possesses unique natural environments for such studies, it often lacks the technical resources and bureaucratic flexibility needed to explore them in depth, particularly in protected, remote, or hard-to-access areas.

But just as important as strengthening scientific capabilities is reinforcing the social dimension of this community. Science cannot be developed in isolation from society. Astrobiology—with its capacity to inspire wonder and its fundamental questions about life, the planet, and the future of the biosphere—holds immense potential to inspire, provoke reflection, and engage the public. The ultimate goal of science is not only to generate knowledge but also to share it and return it to society. This is not merely an ethical responsibility; it is an integral part of scientific work. Researchers funded by public resources have a duty to make their results accessible and to foster scientific literacy by equipping society with the tools needed to understand their environment, make informed decisions, and act responsibly—especially, in this case, with regard to our own planet and its biosphere.

This book is not merely a compilation of experiences; it is also a proposal and an invitation to all of Latin America. This invitation is extended to students, encouraging them to join with enthusiasm in this intellectual adventure of exploring both space and our own planet; to educators, urging them to renew their methodologies and connect with the curiosity of their students; to public administrators, calling on them to invest in knowledge as a driver of development; and to all citizens, inspiring them to take an interest in and claim ownership of scientific knowledge. In a world shaken by multiple crises—climatic, social, and health-related—astrobiology, with its focus on the origins of life and the possibility of its existence beyond Earth, reminds us that life on our planet is a rare and precious phenomenon, one that deserves to be understood and protected. As T. S. Eliot once wrote: "We shall not cease from exploration, and the end of all our exploring will be to arrive where we started and know the place for the first time."

Bogotá, Colombia   Miguel Ángel de Pablo Hernández
April 30, 2025

Contents

Chapter 1
Introduction

María Angélica Leal and David Tovar

Abstract The book Astrobiology and Planetary Sciences in Latin America represents the first major collaborative outcome among scientists from different countries in Latin America, as well as the Latin American Astrobiology Network (Red Latinoamericana de Astrobiología), which was formally established in 2018. Its primary objective is to showcase to the international scientific community the advances in research, education, and public outreach in astrobiology and planetary sciences developed across the region.

The first two chapters provide a global context, highlighting key initiatives in the United States and Europe, as well as the central role of the Centro de Astrobiología (CAB) in Spain as an important reference point for the development of research projects throughout Latin America. The subsequent chapters present national case studies that reconstruct the evolution of these disciplines in various countries, emphasizing the region's cultural richness, unique challenges, and vast scientific potential.

M. A. Leal (✉)
Planetary Sciences and Astrobiology Research Group (GCPA), Universidad Nacional de Colombia/Corporación Científica Laguna, Bogota, Colombia
Biology Department, Faculty of Sciences, Universidad Nacional de Colombia, Bogotá, Colombia
Unit of Geology, Universidad de Alcalá, Alcalá de Henares, Spain
e-mail: maria.leal@corpolaguna.org

D. Tovar
Planetary Sciences and Astrobiology Research Group (GCPA), Universidad Nacional de Colombia/Corporación Científica Laguna, Bogota, Colombia
Unit of Geology, Universidad de Alcalá, Alcalá de Henares, Spain
Geosciences Department, Faculty of Sciences, Universidad Nacional de Colombia, Bogotá, Colombia
e-mail: dftovarr@unal.edu.co

D. Tovar, M. A. Leal (eds.), *Astrobiology and Planetary Sciences in Latin America*, https://doi.org/10.1007/978-3-032-01450-4_1

1

Although most regional initiatives have traditionally been documented in Spanish, this English-language volume aims to elevate these efforts to a global stage. The final chapter also acknowledges the significant contributions of countries such as Uruguay, and Venezuela, through the work of the Latin American Astrobiology Network.

This volume invites scientists, educators, students, and science communicators to join a Latin American community that is working with dedication and excellence toward the advancement of astrobiology and planetary sciences in the region, while also enhancing the international visibility of its important scientific contributions.

Astrobiology and Planetary Sciences in Latin America: Research, Education, Public Outreach, and its Potential for Near-Future Investigations is the first major outcome of the consolidation of the Latin American Astrobiology Network (Red Latinoamericana de Astrobiología). The initial ideas for this network emerged in 2015, driven by founding members from Argentina and Peru. This early initiative led to the organization of the First Latin American Astrobiology Congress (Lima, Peru) in 2016. With the formalization of the network in 2018 during the Second Latin American Astrobiology Congress (Bogota, Colombia), key discussions centered around increasing the visibility of ongoing research projects, educational programs, and science outreach strategies related to astrobiology and planetary sciences across the region.

This book, therefore, has a clear objective: to introduce the international scientific community to the rigorous and committed work of astrobiologists, planetary geologists, astronomers, astrochemists, and other space science researchers throughout Latin America.

The chapters two and three provide a global context for astrobiology, outlining—clearly and concisely—the most representative initiatives in the United States and Europe, though not limited to these regions. Chapter 2 is complemented exceptionally well by Chap. 3, which offers a detailed account of the work of the Centro de Astrobiología (CAB) in Spain—a leading institution whose research in planetary sciences and astrobiology serves as a reference point for Latin America.

The subsequent chapters present research processes, educational initiatives, and science outreach programs developed by different authors, with the goal of reconstructing the history of astrobiology and planetary sciences in several Latin American countries. Each country represented in this extraordinary collaborative effort is a universe unto itself, and its cultural richness and unique challenges illustrate that Latin America holds enormous potential for the continued development of initiatives that promote astrobiology and planetary sciences. The majority of documents recording the activities carried out by countries in the region are written in Spanish. This shared language has served as a bridge for building a broad and solid community around these disciplines.

This book, written in English, further enhances the visibility that has been pursued for many years by researchers in each of the participating countries—who view the use of a language other than their own not as a barrier, but as a valuable opportunity to share their progress in astrobiology and planetary sciences with a global audience. This endeavor is highly significant, as it demonstrates that Latin America is developing scientific initiatives that are on par with those being carried out elsewhere in the world.

This volume seeks to present the most representative research efforts from each participating country, as identified by local authors. At the same time, it acknowledges that new initiatives continue to emerge throughout the region, many of which may find in these pages a source of inspiration and a foundation upon which to build future work in astrobiology and planetary sciences.

Although the editors aimed to include chapters from every country in Latin America, various circumstances prevented the participation of some nations in this first edition. It is our hope that this volume will motivate those countries to join future editions.

Nonetheless, recognizing the importance of countries such as Uruguay, and Venezuela in the historical and recent development of astrobiology and planetary sciences in the region, the final chapter—dedicated to the Latin American Astrobiology Network and its nodes—offers special attention to these nations and their key contributions to the consolidation of these fields in Latin America.

We hope that you, esteemed reader, will find in this book the motivation to join the collective effort being carried out across Latin America. And if this is your first encounter with the subject, we hope this book will serve as a gateway into the practices, research areas, and key developments of astrobiology and planetary sciences. We also hope that your reading experience proves enriching and that you learn something new along the way.

With nothing further to add, we invite you to enjoy this collaborative Latin American effort—a product of years of dedication, and one that will undoubtedly continue to bear fruit in the decades to come.

Chapter 2
State of Astrobiology in the World

Sun Kwok, Masatoshi Ohishi, Muriel Gargaud, Nils G. Holm, Joseph A. Nuth, Jesus Martinez-Frias, and Sergio Pilling

Abstract Astrobiology is a rapidly evolving interdisciplinary science that addresses fundamental questions about the origin, evolution, distribution, and future of life in the Universe. This chapter presents a comprehensive overview of the current global state of astrobiology, highlighting major research directions, organizational structures, and international collaborations that sustain and promote the field. Emphasis is placed on the role of the International Astronomical Union (IAU) and its Commission F3 in fostering research, education, and outreach, particularly in developing countries. Key initiatives include training programs, educational platforms, international symposia, and partnerships with other scientific societies. The chapter also outlines future challenges and opportunities, asserting that astrobiology not only integrates diverse scientific disciplines but also offers a unifying framework for understanding our place in the cosmos.

This chapter is in part based on the report of IAU Commission F3 submitted to the IAU.

S. Kwok (✉)
Laboratory for Space Research, The University of Hong Kong, Hong Kong, China

Department of Earth, Ocean, and Planetary Sciences, University of British Columbia, Vancouver, BC, Canada
e-mail: sunkwok@hku.hk

M. Ohishi
Astronomy Data Center, National Astronomical Observatory of Japan, Mitaka, Tokyo, Japan

M. Gargaud
Laboratoire d'Astrophysique de Bordeaux, Université de Bordeaux, Bordeaux, France

N. G. Holm
Department of Geological Sciences, Stockholm University, Stockholm, Sweden

J. A. Nuth
NASA/GSFC, Solar System Exploration Division, Greenbelt, MD, USA

J. Martinez-Frias
Geosciences Institute, IGEO (CSIC-UCM) Earth Dynamics and Observation (Meteorites and Planetary Geosciences Research Group), Madrid, Spain

S. Pilling
Instituto de Pesquisa and Desenvolvimento, Universidade do Vale do Paraiba-Brazil, São Paulo, Brazil

D. Tovar, M. A. Leal (eds.), *Astrobiology and Planetary Sciences in Latin America*, https://doi.org/10.1007/978-3-032-01450-4_2

2.1 What Is Astrobiology?

Astrobiology, also called bioastronomy or exobiology, is the study of the origin, evolution, and distribution of life in the universe. Research in astrobiology encompasses:

- The origin of the biogenic chemical elements and the search, observations, and analysis of biologically relevant molecules in the circumstellar and interstellar media and in external galaxies (Catling, 2013; Cockell, 2015; Gargaud et al., 2005);
- Study of biomolecules and organic solids in primitive solar system bodies such as comets, asteroids, interplanetary dust particles, meteorites and planetary satellites (Gilmour & Sephton, 2003; Gargaud et al., 2023; Kwok, 2011, 2013);
- The search for extant life, evidence of past life, and evidence of prebiotic chemistry on solar system bodies, including Mars, Europa, Titan and Enceladus (Gargaud et al., 2023; Goldsmith & Owen, 2001);
- The study of the origin, early evolution, and environmental constraints for life on early Earth (Kwok, 2017; Lunine, 2005);
- The detection of extra-solar planets and search for spectroscopic evidence of life, habitability, and/or biological activity on extra-solar planets (Horneck & Rettberg, 2007);
- Historical and philosophical issues linked to the origin and evolution of life on Earth and to its possible presence elsewhere in the Universe (Vakoch, 2013); and
- The search for intelligent signals of extra-terrestrial origin (Shklovskii & Sagan, 1966).

2.2 The Astrobiology Community

Astrobiology is an interdisciplinary subject. Our members come from a variety of backgrounds, including astronomy, biology, chemistry, geology, planetary science, and history of science. We integrate results obtained from space missions to planets, planetary satellites, comets and asteroids with laboratory studies of meteorites, remote astronomical spectroscopic observations of circumstellar and interstellar molecules and solids, protoplanetary disks, and exoplanets (Kolb, 2018). Our membership also includes scientists performing laboratory simulations on the formation of organics in the space environment, and laboratory studies on chemical pathways to life.

Our grand goal is to portray a coherent picture of the synthesis of basic ingredients of life (as we know it) in circumstellar envelopes, interstellar medium, and solar system (as well as other planetary systems). We also study the interrelationships between these findings and the results from studying the early Earth. Our mission is to advance our understanding of the origin of life and the possibility of life elsewhere in the Universe.

2.3 The International Astronomical Union

The International Astronomical Union (IAU) was founded in 1919. Its mission is to promote and safeguard the science of astronomy in all its aspects through international cooperation. Its individual members are professional astronomers from all over the world at the Ph.D. level and beyond, and active in professional research and education in astronomy. The scientific and educational activities of the IAU are organized by its 9 Divisions, 35 Commissions, and 55 Working Groups, covering the full spectrum of Astronomy.

The long-term policy of the IAU is defined by the triennial General Assembly and implemented by the Executive Committee, while the day-to-day operations are directed by the IAU officers. The focal point of its activities is the IAU Secretariat, hosted by the Institut d'Astrophysique de Paris, France.

Astrobiology is a highly interdisciplinary field of research that encompasses (and requires) many different research initiatives, both inside and outside the traditional areas of astronomy. As the main international organization in astronomy, IAU is at the position that can play a major role in research in astrobiology, and to have significant contribution to educational aspects of it. The present IAU Commission F3 Astrobiology was created after the commission reform in 2015.

The Commission's predecessor was Commission 51, which was first formed in 1982 as Bioastronomy: search for extra-terrestrial life, and later renamed Bioastronomy in 2006. Comm. 51 is affiliated with IAU Division F, Planetary Systems and Bioastronomy. As part of the reform of commission structures of the IAU in 2015, The present Commission F3 is under the structure of Division F, with a new name of Astrobiology.

The list of officers and members of IAU Commission F3 is maintained by the International Astronomical Union (IAU) and is updated regularly. Elections are held by the IAU on a triennial basis. The Commission has been led by a succession of presidents representing different countries, with leadership roles transitioning every 3 years. A vice president elected during each term is typically expected to assume the presidency in the following cycle. The Commission has maintained an active online presence to serve as the main interface between its members and the public. This platform offers updates on activities, resources, and opportunities for engagement.

Professional scientists whose research is directly related to any branch of astronomy are eligible to apply for Individual Membership. Candidates are usually nominated by a National Member and reviewed by the IAU Executive Committee. Detailed information on qualifications and application procedures is available through official IAU documentation.

As part of the interdisciplinary nature of IAU Astrobiology Commission, we work closely with:

IAU Div C: Education, Outreach and Heritage, and in particular C1 (Astronomy Education and Development) and C2 (Communicating Astronomy with the Public),

IAU Div F: Planetary Systems and Astrobiology (the main affiliated division), and in particular F2 (Exoplanets and the Solar System).

IAU Div H: Interstellar Matter and Local Universe, and in particular Commission H2 (Astrochemistry).

IAU Astrobiology Commission also serves as an interface between astronomy and other scientific disciplines involved in astrobiological research issues such as chemical sciences, life sciences, and Earth sciences.

2.4 Meetings and Conferences

IAU Commission F3 (and its predecessor C51) has a strong and consistent record of running symposia once every 3 years. Previous symposia were held in Boston (1984), Balaton, Hungary (1987), Val Cenis, France (1990), Santa Cruz, U.S.A. (1993), Capri, Italy (1996), Hawaii, U.S.A. (1999), Hamilton Island, Australia (2002), Reykjavik, Iceland (2004), San Juan, Puerto Rico (2007), Montpellier, France (2011), Nara, Japan (2014), and Coyhaique, Chile (2017).

The following conferences and workshops were organized after the establishment of Commission F3:

- Search for water and life's building blocks in the Universe, Focus Meeting during the IAU General Assembly, 3–5 August 2012, Honolulu, Hawaii, USA
- Search for life: from early Earth to exoplanets, 12–16, December 2016, Qhy Nonh, Vietnam
- Astrobiology training school, 24–25 November, 2017, Santiago, Chile
- Astrobiology 2017, 26 November–1 December, 2017, Coyhaique, Chile. This major event of this commission was attended by 200 participants from 27 countries.
- II Latin American Astrobiology Conference, 23–28 July, 2018, Bogotá, Colombia

In 2011 and 2014, the Commission held two joint meetings with the International Society for the Study of the Origin of Life (ISSOL). These meetings, entitled *ORIGINS 2011* and *ORIGINS 2014*, were held at Montpellier, France (July 3–8, 2011) and Nara, Japan (July 6–11, 2014).

In addition to this regular series of conferences, there were also interdisciplinary meetings such as IAU Symposium 251, *Organic Matter in Space* (2008, Hong Kong), and Special Session 16 *Spectral Phenomena in the Interstellar Medium* at the 2012 IAU General Assembly in Beijing. At the IAU General Assembly in Honolulu, the Commission held a Focus Meeting entitled *Search for Water and Life's Building Blocks in the Universe*. During every general assembly, the Commission has held its business meetings as well as commission meetings often with special themes.

2.5 Links with National Bodies

The IAU Astrobiology Commission coordinates efforts in all research and educational areas in astrobiology at the international level, and has established collaborative programs with other national and international scientific societies with related interests. A partial list is given below:

- The Goddard Center for Astrobiology—http://astrobiology.gsfc.nasa.gov
- Astrobiology at NASA—https://astrobiology.nasa.gov/
- ISSOL-The International Astrobiology Society—http://www.issol.org
- European Astrobiology Network Association—http://www.eana-net.eu/
- Australian Centre for Astrobiology—https://www.unsw.edu.au/research/aca
- SETI Institute—http://www.seti.org
- The Astrobiology Society of Britain—http://www.astrobiologysociety.org
- Italian Astrobiology Society—https://www.astrobiologia.it/
- Societe Francaise d'Exobiologie—http://www.exobiologie.fr
- Centro de Astrobiología—http://www.cab.inta.es
- European Astrobiology Institute—https://europeanastrobiology.eu/
- Spanish Planetology and Astrobiology Network (REDESPA)—http://www.icog.es/redespa/
- Latin American Network on Astrobiology—https://www.astrobiologialatam.org/
- Japan Astrobiology Network—https://www.ls.toyaku.ac.jp/astrobiology-japan/en/index.html

2.6 Education and Outreach

Astrobiology is an interdisciplinary subject which draws from research in astronomy, biology, biochemistry, chemistry, geology, microbiology, physics, and planetary science. It also touches upon the disciplines of history, philosophy and sociology. Education in astrobiology therefore helps students develop the awareness that all sciences are related. Study of astrobiology therefore gives students a broader view of science than traditional science courses and offers them a grand perspective of the frontier of science.

The IAU Astrobiology Commission puts great emphasis on education, especially in developing countries. Its activities include lectures and training for young scientists in different fields to prepare for research in astrobiology, outreach functions, and production of educational shows and displays.

One example the members of the Commission helped produce is the sky show *In Search of Cosmic Life* produced by the Hong Kong Space Museum, which will run between February to November 2018.

The IAU Astrobiology Education and Training Working Group was created in October 2015 in order to coordinate training, education and outreach activities in astrobiology at the international level. Five goals are identified:

- To collect all lectures and conferences in astrobiology which have been recorded during the last 10 years (whatever the language is), to categorize them according to their field and the public concerned (from general public to specialist of the field), and to make them available for free on a website so that people with few financial resources can access them remotely. The platform *Online courses in astrobiology* has been launched in November 2017: http://astrobiovideo.com/en/. So far 3 languages are available: French, English and Spanish.
- To produce handbooks and Massive Open Online Courses (MOOCs) for university students—pooling the individual national efforts that are already in progress over the world. Three MOOCs have been produced: (1) Chris Impey: *Astronomy: exploring Time and Space*. (2) Charley Linneweaver: *Are we alone?* (3) Sun Kwok: *Our Place in the Universe*.
- To develop outreach for the general public and high school teachers. One book has been recently published in French for the general public and will be translated soon in different languages: *La plus grande histoire jamais contée*, Belin Publisher.
- To create an annual international astrobiology training school lasting 1–2 weeks, which would train the young generation in the basics of astrobiology. In 2016 the annual Training School *Rencontres Exobio pour Doctorants/Astrobiology Introductory Course* was recognized by the IAU. Training school organized in March 2018.
  In addition to this annual training schools, we have organized twice a 2-days training schools before one international conference to let the local students (and also the conference PhD students) have a basics course in astrobiology: in Vietnam in 2016, just before the conference *Search for life: from early Earth to exoplanets* and in Santiago, Chile in 2017, just before the C.F3 Astrobiology Conference.
  Lectures of all these training schools are recorded and are available on the online courses in astrobiology.
- To organize a regular international workshop on education in astrobiology in order to discuss how to carry out multidisciplinary training in astrobiology, how to evaluate students, and above all to share all training materials that each country may have developed but kept in its national drawers.

The first International Symposium on Education in Astronomy and Astrobiology (ISE2A) was co-organised in Utrecht in July 2017 in collaboration with the C.C1 Education commission: https://ise2a.uu.nl/. It gathered 100 participants, the next one will be organised in Malaga in 2020.

In the coming years, IAU Astrobiology Education and Training Working Group plans

- To develop further the platform *online courses in astrobiology*, to implement the display of MOOCs and to add lectures in languages other than English, French and Spanish.
- To organize, as often as possible, a 2-days astrobiology training schools before each important international astrobiology conference, in addition to the *Rencontres Exobio pour Doctorants/Astrobiology Introductory Course* which is organised in France in English every year
- To develop outreach for the general public and high school teachers in order to train the trainers, specifically in countries where astrobiology is still under-developed or just developing.

2.7 Encyclopaedia and Monographs

Members of the commission have had major contributions to much advancement in research on the origin of life and astrobiology, and have also played instrumental role in the development of the second edition of the *Encyclopaedia of Astrobiology* was published in 2015 and a continuously updated living edition is now being implemented. This volume of 3000 entries has served as a valuable resource and reference for researchers, teachers, and students.

A *Handbook of Astrobiology*, Vera Kolb (ed.), will be published by CRC Press (Taylor and Francis Group) in 2018.

2.8 Conclusion and Summary

Astrobiology is a rapidly developing interdisciplinary research at the frontier of science. Research in astrobiology is being carried out in many countries in the world. The International Astronomical Union is active in promoting educational efforts in astrobiology, in particular in developing countries outside of North America and Europe. We expect the discipline of astrobiology will continue to grow in the coming decades as new space vehicles and astronomical telescopes explore the Solar System in search for evidence for extra-terrestrial life. The implication of such discovery on human society will be tremendous.

Acknowledgement We thank the IAU for its continued support of astrobiology research and education. This article is in part based on a report submitted by Commission F3 to the IAU in 2018.

Declarations

Competing Interests The authors declare no competing interests.

References

Catling, D. C. (2013). *Astrobiology: A very short introduction*. Oxford University Press. ISBN 978-0199586455.

Cockell, C. S. (2015). *Astrobiology: Understanding life in the universe*. Wiley. ISBN 978-1-118-91332-1.

Gargaud, M., Barbier, B., Martin, H., & Reisse, J. (2005). *Lectures in astrobiology*. Springer. ISBN 978-3-540-26229-9.

Gargaud, M., Irvine, W. M., Amils, R., Claeys, P., Cleaves, H. J., Gerin, M., Rouan, D., Spohn, T., Tirard, S., & Viso, M. (Eds.). (2023). *Encyclopedia of astrobiology*. Springer. ISBN 978-3-662-65092-9.

Gilmour, I., & Sephton, M. A. (2003). *An introduction to astrobiology*. Cambridge University Press. ISBN 978-1107600935.

Goldsmith, D., & Owen, T. (2001). *The search for life in the universe* (3rd ed.). University Science Books. ISBN 1-891389-16-5.

Horneck, G., & Rettberg, P. (Eds.). (2007). *Complete course in astrobiology*. Wiley-VCH. ISBN 978-3-527-61900-9.

Kolb, V. (Ed.). (2018). *A handbook of astrobiology*. CRC Press (Taylor & Francis Group).

Kwok, S. (2011). *Organic matter in the universe*. Wiley-VCH. ISBN 978-3-527-40986-0.

Kwok, S. (2013). *Stardust: The cosmic seeds of life*. Springer. ISBN 978-3-642-32801-5.

Kwok, S. (2017). *Our place in the universe: Understanding fundamental astronomy from ancient discoveries*. Springer. ISBN 978-3-319-54172-3.

Lunine, J. I. (2005). *Astrobiology: A multidisciplinary approach*. Addison-Wesley. ISBN 978-0805380422.

Shklovskii, I. S., & Sagan, C. (1966). *Intelligent life in the universe*. Holden-Day.

Vakoch, D. A. (Ed.). (2013). *Astrobiology, history, and society: Life beyond Earth and the impact of discovery*. Springer. ISBN 978-3-642-35982-8.

Chapter 3
Centro de Astrobiología Over 20 Years of Research in Spain

J. Miguel Mas-Hesse, Olga Prieto-Ballesteros, Francisco Najarro, Eduardo Sebastián, and Ricardo Amils

Abstract The Centro de Astrobiología (CAB) in Spain is an internationally recognized institution dedicated to advancing research in astrobiology through a comprehensive transdisciplinary approach. Founded in 1999 and integrated into NASA's Astrobiology Institute in 2000, CAB has become a key player in exploring the origin, evolution, and distribution of life in the Universe. This chapter presents an in-depth overview of CAB's scientific structure, research departments, and major achievements, ranging from studies in astrophysics, molecular evolution, and planetary habitability to the development of advanced space instrumentation. Emphasis is placed on CAB's leadership in Mars exploration missions (e.g., REMS, TWINS, MEDA, RLS), its robust experimental infrastructure, and its extensive public outreach and education programs. CAB's model illustrates how multidisciplinary and transdisciplinary collaboration can effectively advance the frontiers of astrobiology and contribute to international space science missions and planetary exploration.

3.1 The Center

The origin of the Centro de Astrobiología (CAB) dates back to the proposal submitted to NASA by a group of Spanish and American scientists led by Juan Pérez-Mercader to join the then newly established NASA Astrobiology Institute (NAI) in 1998. Following a thorough analysis and evaluation of the project, and an exchange of official correspondence at the governmental level, the CAB was incorporated into the NAI in April 2000, thereby becoming its first Associate Member outside the United States (Gargaud et al., 2023).

The Spanish Center for Astrobiology was established as a Joint Center, supported by the National Institute for Aerospace Technology (INTA), which

J. M. Mas-Hesse · O. Prieto-Ballesteros · F. Najarro · E. Sebastián · R. Amils (✉)
Centro de Astrobiología (CAB), INTA-CSIC, Torrejón de Ardoz, Madrid, Spain
e-mail: ramils@cbm.csic.es

© The Author(s), under exclusive license to Springer Nature Switzerland AG 2025

D. Tovar, M. A. Leal (eds.), *Astrobiology and Planetary Sciences in Latin America*, https://doi.org/10.1007/978-3-032-01450-4_3

Fig. 3.1 Main headquarters of the Centro de Astrobiología on the INTA campus, Madrid

contributed its expertise in the field of space research, and the Spanish National Research Council (CSIC), with its extensive scientific background. The Government of the Community of Madrid has also supported CAB's activities since its inception (Parro et al., 2020).

The headquarters of the CAB is located on the INTA campus in Torrejón de Ardoz, 20 km northeast of Madrid (Fig. 3.1). It comprises a main building, inaugurated in 2003, and three annexes: a Molecular Ecology laboratory, an Astronomical Observatory equipped with a robotic telescope, and a pavilion housing offices and a classroom. The total built area is approximately 8000 m^2. In 2012, a second building was inaugurated on the campus of the European Space Astronomy Centre (ESAC) of the European Space Agency, also in the vicinity of Madrid.

3.2 Objectives

Centro de Astrobiología was founded with the aim of unraveling the origin of life—how it emerged—both in space (where?) and in time (when?), always considering it as a physicochemical process intrinsically linked to the evolution of the Universe (Fig. 3.2). The astrobiological approach to addressing these questions is based on a multi- and transdisciplinary perspective, which may be summarized as follows:

The first step involves understanding the origin of chemical elements, starting from a primordial Universe that contained only hydrogen, helium, and traces of other elements (Lunine, 2005). The formation and evolution of stars and galaxies gave rise to all the elements we know today, which were dispersed throughout the interstellar medium (e.g. Mateo-Marti et al., 2019). These elements clustered in the space between stars, forming increasingly complex molecular chains, which we can study today using our radio telescopes.

Subsequent generations of stars were no longer formed from pristine material, but from clouds enriched with a wide variety of elements and complex molecules.

Fig. 3.2 Life as a consequence of the evolution of the Universe (Ñito, 2022, personal communication)

These, in turn, gave rise to planetary systems of complex composition (Gebauer et al., 2017).

Astrophysicists at the CAB study these processes and provide information about the boundary conditions applicable to different planetary systems, including the early Earth. Planetary geologists complement these data with information on the properties of planets and the processes that drive their evolution: density, temperature, presence of tectonics and volcanism, existence of a magnetic field and an atmosphere, interactions between rocky layers and fluids, the establishment of cycles of biologically relevant elements, and ultimately, the potential habitability of planets (Molina et al., 2017).

With these boundary conditions, and by understanding the molecular compounds expected throughout the evolution of different planets, biochemists study the processes that may have given rise to life. We know that this process was fully realized at least on Earth, but we suspect that it may have developed in a similar manner in countless environments—whether on planets, moons of giant planets, or asteroids. Likewise, biologists investigate how, once life emerged on Earth, it adapted to changing environmental conditions and evolved toward complexity, under the assumption that similar processes could have occurred on other planets (Cockell, 2015).

The combination of researchers and technologists from such diverse disciplines allows us to conduct not only a multidisciplinary analysis—in which various fields of interest advance in parallel—but also a transdisciplinary one, where scientists from different areas collaborate with one another and with the engineers who develop the required instrumentation.

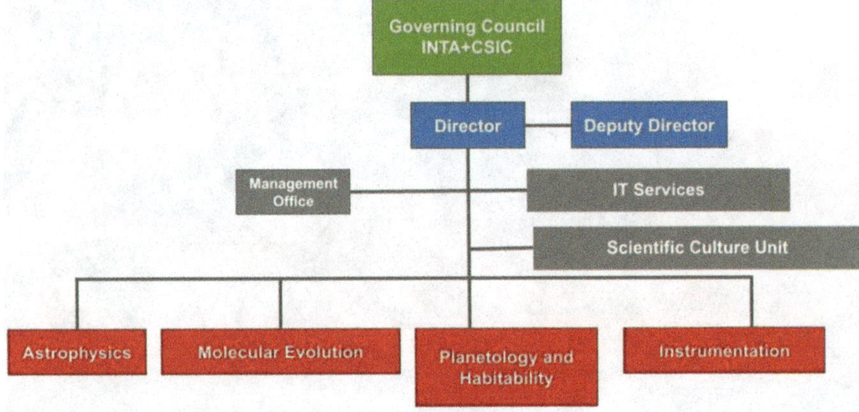

Fig. 3.3 Structure and organization of the Centro de Astrobiología

Finally, engineers design instruments to explore other planets and environments, and they build chambers capable of simulating the conditions of virtually any setting (Wynne et al., 2022).

3.3 Structure and Organization of the Centro de Astrobiología

Currently, the CAB is organized into four departments, each with its corresponding Research Lines (Fig. 3.3):

Department of Astrophysics
- Formation and Evolution of Galaxies
- Interstellar and Circumstellar Medium
- Formation and Evolution of Stars, Brown Dwarfs, and Planets
- Virtual Observatory: Scientific Exploitation of Astronomical Archives

Department of Molecular Evolution
- Prebiotic Chemistry
- Molecular Evolution, RNA World, and Biosensors
- Experimental Evolution Studies with Viruses and Microorganisms
- Microbial Biodiversity
- Molecular Mechanisms of Biological Adaptation
- Biomolecules in Planetary Exploration

Department of Planetology and Habitability
- Planetary Geology and Atmospheres
- Habitability and Extreme Environments

Department of Advanced Instrumentation
* Space Instrumentation
* Simulation Chambers for Planetary Environments

Each of these research lines has a scientific coordinator, whose primary role is to promote scientific activity within the line, as well as to keep the head of the corresponding department informed of the scientific results being achieved and the degree of progress toward the established objectives. Additionally, the CAB is equipped with more than ten laboratories and testing facilities that support all its areas of activity.

3.4 Scientific Activity and Main Results

3.4.1 *Department of Astrophysics*

The activity of the Department of Astrophysics focuses on the study of the processes necessary for the emergence and evolution of life in the Universe. These objectives range from the formation of chemical elements within stars and the formation and evolution of the galaxies that host them, to the processes of planet formation around newly born stars, as well as the formation and evolution of chemical compounds of varying complexity in the interstellar medium (Bujarrabal et al., 2001). Throughout the history of the Universe, successive generations of stars have created all the heavy elements we know within their cores. The atoms of these chemical elements formed molecules, dust grains, and ice mantles in the interstellar and intergalactic medium, eventually condensing into planetary systems with rocky planets rich in water and oxygen, like Earth—conditions under which life emerged over 3 billion years ago, and which are likely to be repeated in countless planetary systems (Lissauer & De Pater, 2013; Najarro et al., 1999).

The department's activity is not limited to the development of scientific programs—whether observational, laboratory-based, or theoretical—but also includes active involvement in the development of astronomical instrumentation, both ground-based and space-based. This instrumentation will make it possible to address current scientific challenges once the various systems become operational in the coming years. Notably, the department is also engaged in activities related to the Virtual Observatory and astronomical databases. The main lines of research are as follows:

3.4.1.1 Evolution of the Universe

Researchers at the Centro de Astrobiología study the formation of the first stars in the history of the Universe. It is believed that this process occurred only 200–300 million years after the Big Bang, when primordial clouds of hydrogen and helium collapsed and triggered the ignition of nuclear fusion processes that produce the energy emitted by stars (Millán-Irigoyen et al., 2021). The ultraviolet radiation emitted by these stars must have ionized the surrounding hydrogen clouds, generating photons associated with the Lyman-alpha transition, which we can now detect and study using sufficiently powerful telescopes (Hayes et al., 2010, 2014). Using the Hubble Space Telescope and the Gran Telescopio Canarias (GTC), we have succeeded in detecting emissions from galaxies within the first billion years of the Universe's history, helping us to understand the initial stages of its evolution. With the launch of NASA and ESA's James Webb Space Telescope (JWST) at the end of 2021, we are now gaining access to the first generations of stars in the history of the Universe, and will be able to verify whether their development unfolded as currently envisioned (De Marco et al., 2022).

3.4.1.2   Star and Exoplanet Formation

Activities in the field of stellar and substellar physics address high-interest scientific objectives that span a broad range of masses—from planets to massive stars of up to 120 solar masses. These include the formation and evolution of such objects, the study of their immediate environments, the characterization of the physicochemical properties of their atmospheres, and the search for (exo)planets. The research also extends to the study of protoplanetary and debris disks (Niedzielski et al., 2021).

3.4.1.3   Astrochemistry in the Interstellar and Circumstellar Medium

Researchers in this group engage in observational, theoretical, and experimental studies of physicochemical processes in interstellar clouds and circumstellar environments. In diffuse clouds, chemical reactions occur, but molecules are generally dissociated due to the intense ultraviolet radiation field. In dense clouds, where young stars are formed, numerous molecular species have been detected as a result of a complex network of chemical reactions and the interaction between the solid phase (dust grains) and the gas phase. Dust grains act as tiny chemical reactors upon which ice is deposited (Kwok, 2013). The irradiation (by photons and ions) of this ice generates species of prebiotic interest, which can be incorporated into comets and other bodies of the early solar system. The vast envelopes surrounding evolved stars—key sites of molecular formation—as well as the chemistry of protoplanetary disks are also under investigation.

3.4.1.4 Virtual Observatory: Scientific Exploitation of Astronomical Archives

The activity of the Data Archive Unit focuses on four main lines of work: the management of astronomical archives; the development of data analysis tools and Virtual Observatory (VO) standards; the implementation of VO science projects; and educational and outreach activities.

Furthermore, the two main space missions currently involving members of the Department of Astrophysics are the following:

PLATO: Scientists at the CAB coordinate the Spanish contribution to the PLATO mission of the European Space Agency, which is scheduled for launch in 2026. Its objective is to detect and characterize (in terms of composition, density, presence of atmospheres, etc.) Earth-like planets orbiting Sun-like stars at similar distances. The interest in finding terrestrial analogs lies in the fact that if life emerged on such planets long ago, it may have evolved in ways similar to life on Earth, making it easier for us to identify from afar (Mateo-Marti et al., 2019). Atmospheric biomarkers defined for Earth are expected to be applicable to these exoplanets as well, whereas in more extreme environments, adaptation may have led to such significant metabolic deviations that their remote detection becomes more challenging. While PLATO instrumentation is being developed, we work with data provided by other space missions, such as NASA's Kepler and TESS, and ESA's Cheops, to search for potentially habitable planets, even if their conditions differ from those on Earth.

BepiColombo/MIXS: In October 2018, the European Space Agency launched the BepiColombo mission to Mercury, where it is expected to arrive in late 2026. The CAB has participated in the development of the MIXS instrument, a telescope capable of performing two-dimensional spectroscopy in the X-ray range (Treis et al., 2008). Mercury is a particularly interesting and relevant planet due to its proximity to the Sun, which has subjected it to intense bombardment by comets and asteroids throughout the history of the Solar System (Bunce et al., 2020). As a result, its surface is deeply altered. In addition to native Mercury material, visible in its large impact craters, its plains contain remnants of the primordial material from which the Solar System was formed. MIXS will enable the mapping of the relative abundances of various elements across different regions of Mercury's surface, helping us to better understand the formation processes of the Solar System—and, by extension, the emergence of life on our own planet.

3.4.2   Department of Planetology and Habitability

The scientific objective of the researchers in this department is to characterize the habitability conditions of the Solar System, both past and present, through the study of information recorded in the geospheres, biospheres, hydrospheres, and atmospheres of terrestrial extreme environments, as well as in other potentially habitable planetary bodies (e.g., Mars, Europa) (Rodriguez-Manfredi et al., 2023; Zorzano et al., 2009; Prieto-Ballesteros et al., 2005; Kargel et al., 2000). The results of analyses on the interaction between environment (geology, physical conditions) and biology are applied in the search for signs of life—both on Earth and beyond our planet (Viúdez-Moreiras et al., 2021). There are two research groups within this department, which interact closely due to the inherently interdisciplinary nature of the objective described.

3.4.2.1   Planetary Geology and Atmospheres

The main activity focuses on the study of planetary bodies in the Universe that possess solid surfaces, including Earth (Lissauer & De Pater, 2013). Research is conducted on the origin, composition, structure, and the processes and agents through which planets, satellites, comets, asteroids, and meteorites have evolved since their formation, with particular emphasis on potentially habitable bodies.

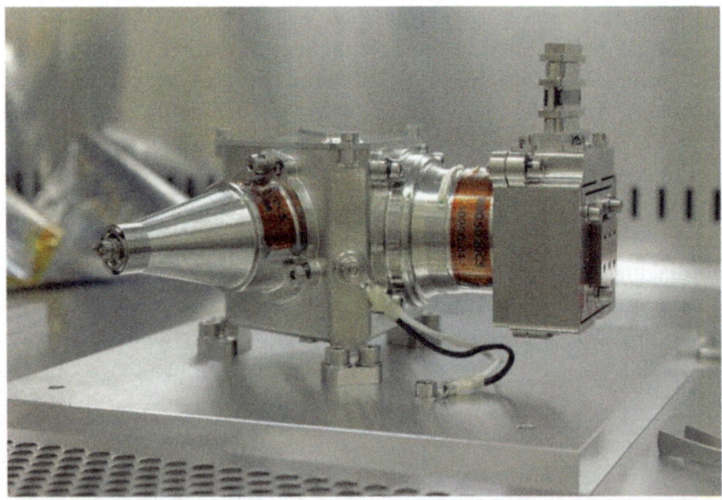

Fig. 3.4 RLS/ExoMars-2020 Instrument. ExoMars 2020 is the first Mars mission led by ESA with a clear astrobiological objective. Its main goals are to search for signs of present and past life on Mars and to geochemically characterize the potentially habitable environment. CAB-INTA has led the development of the Raman spectrometer, which will be the first instrument of its kind to analyze samples from beyond Earth

3.4.2.2 Habitability and Extreme Environments

Researchers in this group aim to characterize extreme environments of astrobiological interest and their biodiversity in order to understand the limits of life and support its search beyond Earth.

Among the key activities carried out by the department are:

- Participation in Solar System exploration missions, at various stages including planning (JUICE, BepiColombo), scientific exploitation (MSL, ExoMars-2016), and instrument development (RLS/ExoMars-2020) (Fig. 3.4).
- Characterization of extreme environments as terrestrial analogs. Extreme environments on Earth are used by CAB researchers to understand the physicochemical and geological processes associated with potential habitability and the presence of life on early Earth and other planets. Through European programs, the study of analogs of great interest to the international scientific community is coordinated, such as Río Tinto (Spain) (Amils et al., 2014), the Tírez Lagoon (Spain), and the Dallol hydrothermal area (Ethiopia) (Fig. 3.5). Field campaigns are also regularly conducted in other Martian and ocean moon analogs, such as Antarctica, hydrothermal regions in Iceland and New Zealand, impact craters in Scandinavia, the Atacama Desert (Chile), and the Chott el Djerid Desert (Tunisia) (Stivaletta et al., 2009). These campaigns often involve international collaborations and are used for testing instrumentation under extreme conditions.
- Simulation and Modeling of Planetary Material and Characterization of Geological Processes.

To this end, the center uses unique simulation chambers—designed by its own researchers—which are capable of recreating surface, atmospheric, and interior

Fig. 3.5 Acidic brines of Dallol (Ethiopia)

Fig. 3.6 EPIC Cannon

environments of planetary bodies, as well as interstellar conditions. These chambers are equipped with radiation sources and various spectroscopic and spectrometric techniques for in situ sample analysis. They have been developed to study physico-chemical and biological processes occurring in extreme environments across the Universe, and to test the performance of space instrumentation. Three of these chambers are particularly noteworthy in relation to the department's activities:

- EPIC (The Experimental Projectile Impact Chamber): A testbed designed to evaluate and support the interpretation of field and remote sensing data related to impact crater formation processes (Fig. 3.6).
- PASC (Planetary Ambient Simulation Chamber): An ultra-high vacuum system capable of precisely reproducing the surface and atmospheric conditions of most planets and moons in the Solar System.
- HPSC (High Pressure Simulation Chamber): A high-pressure cell designed to simulate the deep environments of planetary bodies that possess subsurface aqueous layers.

3.4.3   Department of Molecular Evolution

Life is the result of the evolution of matter and energy in the Universe. In this department, six research groups composed of chemists, physicists, and biologists aim to address various questions related to the origin and evolution of life on our planet, as well as to develop techniques that enable the identification of life on other planetary bodies (Lunine, 2005).

3.4.3.1 Prebiotic Chemistry Group

This group studies critical aspects preceding the origin of life, such as the emergence of biological symmetry, the formation of the first catalytic polymers, and the self-organization of molecules on surfaces. This line of research focuses on prebiotic chemistry, which encompasses all natural physicochemical processes occurring within a planetary environment—from its formation to the emergence of the first self-replicating system in which Darwinian selection processes began to operate.

The group seeks to address several important unresolved questions: (1) the abiotic origin of chemical precursors of life, such as amino acids and simple sugars, and how they combine to form abiotic oligomers and polymers with novel properties (e.g., catalytic, metabolic, etc.); (2) how the homochirality of today's bio-organic compounds originated; (3) how complex molecules form supramolecular assemblies such as vesicles, membranes, etc.; (4) the self-assembly of molecules on surfaces.

3.4.3.2 Molecular Evolution, RNA World, and Biosensors Group

The scientific objectives of this group are as follows: (1) Investigation of heterogeneous prebiotic scenarios and system chemistry-based approaches to the origin and early evolution of life; (2) Study of the polymerization of early-emerging macromolecules that combine genetic information with functional capabilities (including the polymerization of ribonucleotides in catalytic media) and the genotype–phenotype relationships in RNA; (3) Analysis of the three-dimensional folding of RNA molecules under various ionic conditions using biochemical methods, DNA microarrays, and atomic force microscopy (AFM); (4) Optimization of in vitro evolution systems for nucleic acids, including the selection of RNA and DNA aptamers specific to low-molecular-weight biomarkers, RNA molecules, and proteins. Development of aptamers for diagnostic and therapeutic applications in the field of virology; (5) Development of label-free microarrays and biosensors based on nanotechnology, using natural nucleic acids or artificial analogs immobilized on various surfaces as probes.

3.4.3.3 Experimental Evolution Studies with Viruses and Microorganisms Group

This group investigates the dynamics and molecular foundations of biological adaptation through controlled laboratory experiments using viruses and bacteria. The main objectives are: (1) Study of the relationships between mutation rate, adaptive capacity, and viral population extinction; (2) Analysis of the evolutionary pathways followed by microbial populations exposed to different patterns of selective pressure changes; (3) Reconstruction of fitness landscapes based on real data obtained through high-throughput sequencing techniques; and (4) Study of the influence of external environmental agents on the stability and evolutionary potential of acellular replicators.

3.4.3.4 Molecular Mechanisms of Biological Adaptation Group

This group focuses on studying molecular strategies and metabolic functions that enable microorganisms from natural samples to adapt to extreme conditions. To achieve this, molecular biology and omics techniques such as functional metagenomics and metatranscriptomics are used, allowing access to and study of all microorganisms within a community, whether cultivable or not.

Another research line involves studying the transfer of genetic material within bacterial populations via extracellular DNA and its relevance in adaptation. This group has identified novel resistance mechanisms to: (1) arsenic, nickel, and acidic pH in planktonic and rhizosphere microorganisms from the Río Tinto; (2) sodium chloride and radiation in microorganisms from various hypersaline environments (Andean hypersaline lakes, salt flats, and rhizospheres of halophytic plants).

Additional objectives include: (1) searching for cold-resistance genes in rhizospheric microorganisms of Antarctic plants; (2) developing plants with resistance genes to acidic pH, UV radiation, and NaCl; (3) implementing functional metagenomics using the development of hyperhalophilic vectors and hosts for gene expression from hypersaline environments, and employing microfluidics to screen metagenomic libraries.

3.4.3.5 Microbial Biodiversity Group

This group studies microorganisms and microbial populations inhabiting terrestrial environments analogous to potentially habitable extraterrestrial regions. Specifically, it focuses on the biodiversity, ecology, and evolution of these organisms. In addition, it investigates the limits of life and the molecular and cellular mechanisms that enable adaptation to extreme environments. The research approach combines fieldwork—dedicated to sample collection and environmental parameter assessment—with laboratory-based experimental studies.

The objectives of this research group are as follows: (1) Description of microbial ecology and the development of models to explain ecological patterns and environmental adaptation; (2) Identification of essential genes/functions involved in adaptation using metatranscriptomics, metabolomics, and metaproteomics; (3) Analysis of key stress factors affecting the metabolic activity of major microbial communities.

These objectives are addressed through two main lines of work: (1) Diversity, distribution, and adaptation of psychrophilic microorganisms (Antarctica, glaciers) and extreme acidophiles (studies on adaptation to toxic metals and radiation in acidophilic algae from Río Tinto, and microbial diversity in other acidic environments) (e.g. Amils et al., 2014); (2) Microbial ecology of the atmosphere.

3.4.3.6   Biomolecules in Planetary Exploration Group

This group aims to understand the metabolic potential and the preservation of molecular biomarkers across space and time in environments analogous to those identified on other planetary bodies. Its work adopts a transdisciplinary approach to provide an integrated view of planetary habitability.

The experimental methodology is based on the in situ study of biological material (through field campaigns in saline deserts, cold deserts, Arctic and Antarctic permafrost, and the deep subsurface), its preservation and interaction with the physical environment, the identification of molecular biomarkers, and the development of methodologies and instruments for their in situ detection.

This group has led the design of the SOLID (Signs Of Life Detector) instrument, which enables the detection and identification of microorganisms and biochemical compounds through the automated analysis of solid samples (soil, rock, or ice) and liquid samples. At the core of SOLID is a biochip containing over 300 antibodies, known as LDChip (Life Detector Chip), capable of detecting a similar number of compounds or microorganisms. Currently, SOLID has reached a Technology Readiness Level (TRL) 6, according to NASA standards, and has been proposed in several calls for instrumentation for Mars exploration.

3.4.4   Department of Instrumentation

The Instrumentation Department at the CAB conducts all its research activities in the field of space instrumentation technologies for planetary exploration and astrophysics. These tasks arise from a multi- and transdisciplinary collaboration between members of the Instrumentation Department and scientists across the center (e.g. Farley et al., 2020).

The department's technological scope ranges from the conception of prototypes and new instrumentation concepts to the design and supervision of specialized industry for the manufacturing of flight models. This broad development process often includes testing campaigns to validate and mature the developed instrumentation and technologies, either in simulation chambers or in environmentally representative settings (so-called terrestrial analogs) that replicate parameters such as temperature, pressure, wind, aqueous environments, or extreme acidity.

To support this, the department is equipped with various infrastructures, including planetary simulation chambers and vacuum technologies housed in the Laboratory for Planetary Environment Simulation at CAB, providing scientific and technological support to the center's research community.

Within the department's framework, operations and exploitation of scientific data collected by its technological developments are also carried out, in close collaboration with other staff and associated research units.

Following an initial phase focused on prototype and field model design, since 2005 the department has consolidated the experience gained in previous projects and redirected its efforts toward the development of flight instrumentation. To that end, it has established strong collaborative ties with both national and international research centers as well as with the Spanish aerospace industry.

Notably, the department has taken a leading role in the development of environmental monitoring stations for Mars exploration, such as REMS, TWINS, and MEDA (Rodriguez-Manfredi et al., 2021; Sebastián et al., 2021). It has also collaborated on other flight instruments such as the RLS–Raman Laser Spectrometer for ESA's ExoMars mission, in partnership with the associated UVA–CSIC unit, as

Fig. 3.7 Photograph of Boom 2 and the UV sensor of REMS, taken by one of Curiosity's cameras in a Martian dune landscape. The air temperature sensor, humidity sensor, and wind sensor can be seen on Boom 2

Fig. 3.8 The InSight mission on the surface of Mars. One of the TWINS booms can be seen in the image

well as on the PLATO telescope and the SAFARI/SPICA spectrometer with the Department of Astrophysics.

The REMS (Rover Environmental Monitoring Station) instrument, onboard NASA's Mars Science Laboratory rover Curiosity, has been monitoring and characterizing the environment of Gale Crater on Mars since August 2012. REMS continuously collects data on atmospheric and ground temperature, pressure, wind speed and direction, atmospheric humidity, and incident ultraviolet radiation (Ruíz et al., 2025) (Fig. 3.7). Over this period, the instrument has recorded more than 30 million readings from each of its sensors. Members of the department are actively involved in the scientific analysis of this data and regularly participate in scientific discussions and meetings held with national and international members of the REMS team (Ruíz et al., 2025; Vasavada et al., 2017).

TWINS (Temperature and Wind Sensors for the InSight Mission) is an instrument composed of sensors for wind direction, wind speed, and atmospheric temperature, integrated into the lander of NASA's *InSight* mission, which is dedicated to characterizing the interior of the planet Mars and was launched in May 2018 (Spiga et al., 2018) (Fig. 3.8). TWINS makes use of spare models originally

Fig. 3.9 One of the MEDA booms mounted on the Perseverance rover on the surface of Mars

developed for the REMS instrument, which were reconfigured and adapted to meet the specific requirements and constraints of the *InSight* mission.

The scientific role and contributions of the TWINS instrument to the *InSight* mission were of vital importance:

1. In the initial phase, it characterized the mission's environmental conditions, determining the precise moment for carrying out the critical deployment of onboard instruments—specifically, when wind and other environmental factors were sufficiently calm;
2. Once the instruments had been deployed, the scientific data provided by TWINS enabled the filtering out of false seismic readings caused by strong ambient winds;
3. Finally, the high-resolution data collected by TWINS proved to be of great scientific value, complementing those obtained by REMS in Gale Crater, and together forming the first meteorological–environmental network on the surface of Mars (Vasavada et al., 2017).

The MEDA (Mars Environmental Dynamics Analyzer) instrument, developed for NASA's Mars 2020 mission, represents the natural evolution of REMS in terms of performance and the range of meteorological variables measured (Rodriguez-Manfredi et al., 2021) (Fig. 3.9). The instrument was launched in 2020 following the assembly, calibration, and acceptance of its various flight models by the research team.

Its primary scientific objective is to characterize local environmental parameters, the physical properties of dust, and to study habitability conditions for future human exploration (Sebastián et al., 2021; Zorzano et al., 2009). To achieve this, the instrument was designed as a suite of sensors that measure:

- Relative humidity (Relative Humidity Sensor—HS),
- Air temperature (Air Temperature Sensor—ATS),
- Net infrared radiation balance (Thermal IR Sensor—TIRS),

- Wind speed and direction on Mars (Wind Sensors—WS),
- Radiation and properties of suspended aerosols (Radiation and Dust Sensor RDS), and
- Atmospheric pressure (Pressure Sensor—PS) (Rodriguez-Manfredi et al., 2023).

3.4.5 *Unit of Scientific Culture*

Given the high level of public interest in questions related to Astrobiology, the CAB has prioritized outreach, dissemination, and science communication activities since its foundation, with the aim of:

- Ensuring the transfer of knowledge across different disciplines, as well as with the industrial sector;
- Training a new generation of astrobiologists;
- Promoting the spread of scientific culture in Spain and other countries.

The Scientific Culture Unit of the CAB (UCC-CAB), under the authority of the Center's Directorate, facilitates interaction between researchers and the general public, making astrobiology a science accessible to all. The general objectives of the UCC-CAB are to promote the dissemination of scientific advances in the field of astrobiology, to optimize the internal and external communication channels available to the CAB, and to enable smooth interaction between the CAB's research community and the broader public.

Since its creation, the CAB has worked to foster a scientific culture rooted in these pillars, integrating into the UCC-CAB all actions and policies related to knowledge dissemination, public relations, institutional communication, and the overall representation and image of the Center.

In schematic form, the lines of action of the UCC-CAB are as follows, addressing a range of different areas:

Targeted at the Scientific and Academic Community
- Seminars and colloquia held in the CAB Auditorium
- International School of Astrobiology *Josep Comas i Solà*

Targeted at the Educational Community and the General Public
- Guided school visits to the CAB
- Seminars and lectures for school students and the general public

Organization of workshops and talks at science outreach events for students and the general public, including:

- Science Week (Community of Madrid)
- Science Fairs (Nationwide in Spain)
- Scientific Weekend (Finde Científico, Community of Madrid)
- Ciudad Ciencia (Nationwide in Spain)
- European Researchers' Night (Community of Madrid)

- Ciencia en el Barrio (Community of Madrid)
- Expoastronómica (Community of Castilla-La Mancha)

Educational Programs

- PARTNER (INTA–NASA): An academic project involving NASA's radio telescope in Robledo, focused on educational radio astronomy for students.
- CESAR Project (INTA–ESA–Isdefe): *Cooperation through Education in Science and Astronomy Research*, aimed at teaching astronomy, planetary sciences, and space exploration to school students.
- MeteoMars Project: A Mars meteorology initiative designed for school settings.
- Didactic Applications of the Virtual Observatory: Educational use of Virtual Observatory tools in the classroom and in citizen science activities.

Targeted at the Media

- Preparation of press releases and media engagement
- *Astrobiology Series* (CEMAV–UNED TV)

Other Areas of Action

- Documentation and library services
- Editorial coordination

  - Publication of Zoé, Journal of Astrobiology
  - Technical support and promotion of outreach and institutional events

- CAB Website
- Science outreach blogs
- Facebook and Twitter (X)
- CAB YouTube channel

Attendance at conferences on science communication, education, and outreach The CAB also hosts the Spanish coordination office of the European Southern Observatory's Outreach Network (ESON). The CAB's Scientific Culture Unit is a member of the Network of Scientific Culture and Innovation Units (Red UCC+i) of the Spanish Foundation for Science and Technology (FECYT).

Declarations

Competing Interests The authors declare no competing interests.

References

Amils, R., Fernández-Remolar, D., & Ipbsl Team. (2014). Río Tinto: A geochemical and mineralogical terrestrial analogue of Mars. *Life, 4*(3), 511–534. https://doi.org/10.3390/life4030511

Bujarrabal, V., Castro-Carrizo, A., Alcolea, J., & Contreras, C. S. (2001). Mass, linear momentum and kinetic energy of bipolar flows in protoplanetary nebulae. *Astronomy & Astrophysics, 377*(3), 868–897.

Bunce, E. J., Martindale, A., Lindsay, S., Muinonen, K., Rothery, D. A., Pearson, J., et al. (2020). The BepiColombo mercury imaging X-ray spectrometer: Science goals, instrument performance and operations. *Space Science Reviews, 216,* 1–38. https://doi.org/10.1007/s11214-020-00750-2

Cockell, C. S. (2015). *Astrobiology: Understanding life in the universe.* Wiley. ISBN 978-1-118-91332-1.

De Marco, O., Akashi, M., Akras, S., Alcolea, J., Aleman, I., Amram, P., Balick, B., De Beck, E., Blackman, E. G., Boffin, H. M. J., Boumis, P., Bublitz, J., Bucciarelli, B., Bujarrabal, V., Cami, J., Chornay, N., Chu, Y.-H., Corradi, R. L. M., Frank, A., García-Hernández, D. A., García-Rojas, J., García-Segura, G., Gómez-Llanos, V., Gonçalves, D. R., Guerrero, M. A., Jones, D., Karakas, A. I., Kastner, J. H., Kwok, S., Lykou, F., Manchado, A., Matsuura, M., McDonald, I., Miszalski, B., Mohamed, S. S., Monreal-Ibero, A., Monteiro, H., Montez, R., Moraga Baez, P., Morisset, C., Nordhaus, J., Mendes de Oliveira, C., Osborn, Z., Otsuka, M., Parker, Q. A., Peeters, E., Quint, B. C., Quintana-Lacaci, G., Redman, M., Ruiter, A. J., Sabin, L., Sahai, R., Sánchez Contreras, C., Santander-García, M., Seitenzahl, I., Soker, N., Speck, A. K., Stanghellini, L., Steffen, W., Toalá, J. A., Ueta, T., Van de Steene, G., Van Winckel, H., Ventura, P., Villaver, E., Vlemmings, W., Walsh, J. R., Wesson, R., & Zijlstra, A. A. (2022). The messy death of a multiple star system and the resulting planetary nebula as observed by JWST. *Nature Astronomy, 6*(12), 1421–1432. https://doi.org/10.1038/s41550-022-01845-2

Farley, K. A., Williford, K. H., Stack, K. M., Bhartia, R., Chen, A., de la Torre, M., Hand, K., Goreva, Y., Herd, C. D. K., Hueso, R., Liu, Y., Maki, J. N., Martinez, G., Moeller, R. C., Nelessen, A., Newman, C. E., Nunes, D., Ponce, A., Spanovich, N., Willis, P. A., Beegle, L. W., Bell, J. F., III, Brown, A. J., Hamran, S.-E., Hurowitz, J. A., Maurice, S., Paige, D. A., Rodriguez-Manfredi, J. A., Schulte, M., & Wiens, R. C. (2020). Mars 2020 mission overview. *Space Science Reviews, 216,* 1–41. https://doi.org/10.1007/s11214-020-00762-y

Gargaud, M., Irvine, W. M., Amils, R., Claeys, P., Cleaves, H. J., Gerin, M., Rouan, D., Spohn, T., & Tirard, S. (2023). & Viso. In M. (Ed.), *Encyclopedia of astrobiology.* Springer. ISBN 978-3-662-65092-9.

Gebauer, S., Grenfell, J. L., Stock, J. W., Lehmann, R., Godolt, M., von Paris, P., & Rauer, H. (2017). Evolution of earth-like extrasolar planetary atmospheres: Assessing the atmospheres and biospheres of early earth analog planets with a coupled atmosphere biogeochemical model. *Astrobiology, 17*(1), 27–54. https://doi.org/10.1089/ast.2015.1384

Hayes, M., Östlin, G., Schaerer, D., Mas-Hesse, J. M., Leitherer, C., Atek, H., Kunth, D., Verhamme, A., de Barros, S., & Melinder, J. (2010). Escape of about five per cent of Lyman-α photons from high-redshift star-forming galaxies. *Nature, 464*(7288), 562–565. https://doi.org/10.1038/nature08881

Hayes, M., Östlin, G., Duval, F., Sandberg, A., Guaita, L., Melinder, J., Adamo, A., Schaerer, D., Verhamme, A., Orlitová, I., Mas-Hesse, J. M., Cannon, J. M., Atek, H., Kunth, D., Laursen, P., Otí-Floranes, H., Pardy, S., Rivera-Thorsen, T., & Herenz, E. C. (2014). The Lyman alpha reference sample. II. Hubble Space Telescope imaging results, integrated properties, and trends. *The Astrophysical Journal, 782*(1), 6. https://doi.org/10.1088/0004-637X/782/1/6

Kargel, J. S., Kaye, J. Z., Head, J. W., III, Marion, G. M., Sassen, R., Crowley, J. K., Prieto Ballesteros, O., Grant, S. A., & Hogenboom, D. L. (2000). Europa's crust and ocean: Origin, composition, and the prospects for life. *Icarus, 148*(1), 226–265. https://doi.org/10.1006/icar.2000.6471

Kwok, S. (2013). *Stardust: The cosmic seeds of life.* Springer. ISBN 978-3-642-32801-5.

Lissauer, J. J., & De Pater, I. (2013). *Fundamental planetary science: Physics, chemistry and habitability.* Cambridge University Press.

Lunine, J. I. (2005). *Astrobiology: A multidisciplinary approach.* Addison-Wesley. ISBN 978-0805380422.

Mateo-Marti, E., Prieto-Ballesteros, O., Munoz Caro, G., González-Díaz, C., Muñoz-Iglesias, V., & Gálvez-Martínez, S. (2019). Characterizing interstellar medium, planetary surface and deep environments by spectroscopic techniques using unique simulation chambers at Centro de Astrobiologia (CAB). *Life, 9*(3), 72. https://doi.org/10.3390/life9030072

Millán-Irigoyen, I., Mollá, M., Cerviño, M., Ascasibar, Y., García-Vargas, M. L., & Coelho, P. R. T. (2021). HR-pypopstar: High-wavelength-resolution stellar populations evolutionary synthesis model. *Monthly Notices of the Royal Astronomical Society, 506*(4), 4781–4799. https://doi.org/10.1093/mnras/stab1969

Molina, A., López, I., Prieto-Ballesteros, O., Fernández-Remolar, D., de Pablo, M. Á., & Gómez, F. (2017). Coogoon Valles, western Arabia Terra: Hydrological evolution of a complex Martian channel system. *Icarus, 293*, 27–44. https://doi.org/10.1016/j.icarus.2017.04.002

Najarro, F., Hillier, D. J., & Figer, D. F. (1999). Metal abundances in the Galactic Center. In *Symposium-international astronomical union* (Vol. 193, pp. 491–492). Cambridge University Press. https://doi.org/10.1017/S007418090020613X

Niedzielski, A., Villaver, E., Adamów, M., Kowalik, K., Wolszczan, A., & Maciejewski, G. (2021). Tracking Advanced Planetary Systems (TAPAS) with HARPS-N-VII. Elder suns with low-mass companions. *Astronomy & Astrophysics, 648*, A58. https://doi.org/10.1051/0004-6361/202037892

Parro, V., Mas-Hesse, J. M., Gomez-Elvira, J., Giménez, Á., & Pérez-Mercader, J. (2020). Introduction—Centro de Astrobiología: 20 years building astrobiology. *Astrobiology, 20*(9), 1025–1028. https://doi.org/10.1089/ast.2020.0804

Prieto-Ballesteros, O., Kargel, J. S., Fernández-Sampedro, M., Selsis, F., Martínez, E. S., & Hogenboom, D. L. (2005). Evaluation of the possible presence of clathrate hydrates in Europa's icy shell or seafloor. *Icarus, 177*(2), 491–505. https://doi.org/10.1016/j.icarus.2005.02.021

Rodriguez-Manfredi, J. A., de la Torre Juárez, M., Alonso, A., Apéstigue, V., Arruego, I., Atienza, T., Banfield, D., Boland, J., Carrera, M. A., Castañer, L., Ceballos, J., Chen-Chen, H., Cobos, A., Conrad, P. G., Cordoba, E., del Río-Gaztelurrutia, T., de Vicente-Retortillo, A., Domínguez-Pumar, M., Espejo, S., Fairen, A. G., Fernández-Palma, A., Ferrándiz, R., Ferri, F., Fischer, E., García-Manchado, A., García-Villadangos, M., Genzer, M., Giménez, S., Gómez-Elvira, J., Gómez, F., Guzewich, S. D., Harri, A.-M., Hernández, C. D., Hieta, M., Hueso, R., Jaakonaho, I., Jiménez, J. J., Jiménez, V., Larman, A., Leiter, R., Lepinette, A., Lemmon, M. T., López, G., Madsen, S. N., Mäkinen, T., Marín, M., Martín-Soler, J., Martínez, G., Molina, A., Mora-Sotomayor, L., Moreno-Álvarez, J. F., Navarro, S., Newman, C. E., Ortega, C., Parrondo, M. C., Peinado, V., Peña, A., Pérez-Grande, I., Pérez-Hoyos, S., Pla-García, J., Polkko, J., Postigo, M., Prieto-Ballesteros, O., Rafkin, S. C. R., Ramos, M., Richardson, M. I., Romeral, J., Romero, C., Runyon, K. D., Saiz-Lopez, A., Sánchez-Lavega, A., Sard, I., Schofield, J. T., Sebastian, E., Smith, M. D., Sullivan, R. J., Tamppari, L. K., Thompson, A. D., Toledo, D., Torrero, F., Torres, J., Urquí, R., Velasco, T., Viúdez-Moreiras, D., Zurita, S., & The MEDA team. (2021). The mars environmental dynamics analyzer, MEDA. A suite of environmental sensors for the mars 2020 mission. *Space Science Reviews, 217*, 48. https://doi.org/10.1007/s11214-021-00816-9

Rodriguez-Manfredi, J. A., de la Torre Juarez, M., Sanchez-Lavega, A., Hueso, R., Martinez, G., Lemmon, M. T., Newman, C. E., Munguira, A., Hieta, M., Tamppari, L. K., Polkko, J., Toledo, D., Sebastian, E., Smith, M. D., Jaakonaho, I., Genzer, M., De Vicente-Retortillo, A., Viudez-Moreiras, D., Ramos, M., Saiz-Lopez, A., Lepinette, A., Wolff, M., Sullivan, R. J., Gomez-Elvira, J., Apestigue, V., Conrad, P. G., Del Rio-Gaztelurrutia, T., Murdoch, N., Arruego, I., Banfield, D., Boland, J., Brown, A. J., Ceballos, J., Dominguez-Pumar, M., Espejo, S., Fairén, A. G., Ferrandiz, R., Fischer, E., Garcia-Villadangos, M., Gimenez, S., Gomez-Gomez, F., Guzewich, S. D., Harri, A.-M., Jimenez, J. J., Jimenez, V., Makinen, T., Marin, M., Martin, C., Martin-Soler, J., Molina, A., Mora-Sotomayor, L., Navarro, S., Peinado, V., Perez-Grande, I., Pla-Garcia, J., Postigo, M., Prieto-Ballesteros, O., Rafkin, S. C. R., Richardson, M. I., Romeral, J., Romero, C., Savijärvi, H., Schofield, J. T., Torres, J., Urqui, R., Zurita, S., & the MEDA team. (2023). The diverse meteorology of Jezero crater over the first 250 sols of Perseverance on Mars. *Nature Geoscience, 16*(1), 19–28. https://doi.org/10.1038/s41561-022-01084-0

Ruíz, M., Sebastián-Martínez, E., Rodríguez-Manfredi, J. A., Pla-García, J., de la Torre-Juarez, M., & Rafkin, S. C. (2025). Meteorological changes across curiosity rover's traverse using REMS measurements and comparisons with measurements and MRAMS model results. *Remote Sensing, 17*(3), 368. https://doi.org/10.3390/rs17030368

Sebastián, E., Martínez, G., Ramos, M., Perez-Grande, I., Sobrado, J., & Manfredi, J. A. R. (2021). Thermal calibration of the MEDA-TIRS radiometer onboard NASA's Perseverance rover. *Acta Astronautica, 182*, 144–159. https://doi.org/10.1016/j.actaastro.2021.02.006

Spiga, A., Banfield, D., Teanby, N. A., Forget, F., Lucas, A., Kenda, B., Rodriguez Manfredi, J. A., Widmer-Schnidrig, R., Murdoch, N., Lemmon, M. T., Garcia, R. F., Martire, L., Karatekin, Ö., Le Maistre, S., Van Hove, B., Dehant, V., Lognonné, P., Mueller, N., Lorenz, R., Mimoun, D., Rodriguez, S., Beucler, É., Daubar, I., Golombek, M. P., Bertrand, T., Nishikawa, Y., Millour, E., Rolland, L., Brissaud, Q., Kawamura, T., Mocquet, A., Martin, R., Clinton, J., Stutzmann, É., Spohn, T., Smrekar, S., & Banerdt, W. B. (2018). Atmospheric science with InSight. *Space Science Reviews, 214*, 1–64. https://doi.org/10.1007/s11214-018-0543-0

Stivaletta, N., Barbieri, R., Picard, C., & Bosco, M. (2009). Astrobiological significance of the sabkha life and environments of southern Tunisia. *Planetary and Space Science, 57*(5-6), 597–605. https://doi.org/10.1016/j.pss.2008.10.002

Treis, J., Hälker, O., Andricek, L., Herrmann, S., Heinzinger, K., Lauf, T., Lechner, P., Lutz, G., Mas-Hesse, J. M., Porro, M., Richter, R. H., San Juan, J. L., Schaller, G., Schnecke, M., Schopper, F., Segneri, G., Soltau, H., Stevenson, T., Strüder, L., & Whitford, C. (2008). DEPFET based X-ray detectors for the MIXS focal plane on BepiColombo. In *High energy, optical, and infrared detectors for astronomy III* (Vol. 7021, pp. 325–336). SPIE. https://doi.org/10.1117/12.787582

Vasavada, A. R., Piqueux, S., Lewis, K. W., Lemmon, M. T., & Smith, M. D. (2017). Thermophysical properties along Curiosity's traverse in Gale crater, Mars, derived from the REMS ground temperature sensor. *Icarus, 284*, 372–386. https://doi.org/10.1016/j.icarus.2016.11.035

Viúdez-Moreiras, D., Richardson, M. I., & Newman, C. E. (2021). Constraints on emission source locations of methane detected by Mars Science Laboratory. *Journal of Geophysical Research: Planets, 126*(12), e2021JE006958. https://doi.org/10.1029/2021JE006958

Wynne, J. J., Titus, T. N., Agha-Mohammadi, A. A., Azua-Bustos, A., Boston, P. J., de León, P., et al. (2022). Fundamental science and engineering questions in planetary cave exploration. *Journal of Geophysical Research: Planets, 127*(11), e2022JE007194. https://doi.org/10.1029/2022JE007194

Zorzano, M. P., Mateo-Martí, E., Prieto-Ballesteros, O., Osuna, S., & Renno, N. (2009). Stability of liquid saline water on present day Mars. *Geophysical Research Letters, 36*(20). https://doi.org/10.1029/2009GL040315

Chapter 4
Brief History of Astrobiology in Argentina

Ximena C. Abrevaya

Abstract Astrobiology is a scientific field defined as the study of the origin, evolution, and distribution of life in the universe. In Argentina, around the 1960s, early formal attempts to conduct scientific research in the "search for signals from technologically advanced extraterrestrial civilizations" or "intelligent extraterrestrial life" emerged, commonly associated with the SETI project (Search for Extraterrestrial Intelligence). This led to the development of a unique SETI project in the Southern hemisphere, the only one of its kind in Latin America.

Later, the search for life in the universe took a different direction in the country with the rise of new studies on planetary habitability. These efforts, rooted in stellar astrophysics, later incorporated approaches from the biological sciences. This innovation sparked interdisciplinary studies and included laboratory experiments in this developing field.

The initial interdisciplinary studies were expanded with contributions from other areas such as geology and cosmochemistry, broadening the scope of existing research topics. As a result of collaborative efforts among Argentinian researchers from various disciplines, the "Argentinian Research Unit in Astrobiology" (Astrobio. ar) was established. The theme of the original investigations into the search for intelligent extraterrestrial life was revisited from a new perspective with the OTHER project.

This chapter provides a brief overview of the key events that led to the development of astrobiology in Argentina.

X. C. Abrevaya (✉)
Instituto de Astronomía y Física del Espacio, UBA-CONICET,
Ciudad Autónoma de Buenos Aires, Argentina

Facultad de Ciencias Exactas y Naturales, Universidad de Buenos Aires,
Ciudad Autónoma de Buenos Aires, Argentina

Argentinian Research Unit in Astrobiology (Astrobio.ar),
Ciudad Autónoma de Buenos Aires, Argentina
e-mail: abrevaya@iafe.uba.ar

35

D. Tovar, M. A. Leal (eds.), *Astrobiology and Planetary Sciences in Latin America*, https://doi.org/10.1007/978-3-032-01450-4_4

4.1 Introduction

Astrobiology is a relatively new area of science that explores the possibilities of life existing beyond Earth, as well as the origin and evolution of life on our planet. It can be defined strictly as "the study of the origin, evolution, and distribution of life in the universe." This field not only involves the exploration of planetary bodies close to or distant from Earth but also the study of our own planet, since, to date, terrestrial life forms are the only known examples of life. Therefore, understanding the mechanisms of the origin and evolution of life on Earth is essential for predicting the existence of life on other planetary bodies.

Several questions coexist that astrobiology seeks to answer, covering various aspects. Some fundamental inquiries include: What is life? How did life originate on Earth? Is it possible to find life or signs of life on other planets? How can we detect the existence of life in the universe? Due to the complexity of finding answers to these questions, astrobiology draws from diverse scientific fields such as physics, chemistry, astronomy, geology, and even philosophy, making it an interdisciplinary or multidisciplinary area. Moreover, some of these inquiries were posed in ancient times, predating astrobiology as a formal science. In fact, the ideas regarding the possibility of other inhabited "worlds" are not new and are referred to as "cosmic pluralism." This possibility of finding life on other planetary bodies was initially raised in ancient Greece and in other Eastern and Western civilizations throughout different historical periods.

Today, we can distinguish two major areas within the exploration of the universe in search of life. On one hand, there is the search for "signals from technologically advanced civilizations" or "intelligent extraterrestrial life," exemplified by what is known as the "SETI project" (Search for Extraterrestrial Intelligence). On the other hand, there is the broader search for life, typically focused on the search for simple life forms (such as microorganisms), which is an inquiry from a biological perspective—what we might call astrobiology in the strict sense of the term. Both approaches employ different strategies and methodologies. For instance, the SETI project employs approaches specifically designed to detect radio or light signals from extraterrestrial civilizations, referred to as "technosignatures." In contrast, the search for life in a generic sense involves a variety of experimental approaches and methodologies, which, paralleling SETI, may include the potential detection of biological signals on other planetary bodies, known as "biosignatures." In 2015, NASA, in its Astrobiology Strategy document, stated that "traditional SETI is not part of astrobiology" (Hays et al., 2015). However, not all of the scientific community agrees with this division, and thus SETI is often included as part of astrobiology (Wright, 2018).

Historically, the SETI project consolidated globally starting in the 1960s, building on its immediate predecessors, the Ozma and "CETI" (Communication with Extraterrestrial Intelligence) projects, although the ideas of establishing contact with extraterrestrial civilizations date back to the early twentieth century. Astrobiology in the strict sense has a more challenging beginning to delineate, as its field of study is much broader and the topics it encompasses have been included, at least fragmentarily, throughout history as part of other areas. Nevertheless, the term astrobiology began to be widely used in the 1990s with the meaning it holds today to define a branch of science, partially replacing the term "exobiology" that described studies related to extraterrestrial life. Although "astrobiology" and "exobiology" are currently used interchangeably, some interpretations consider that exobiology focuses exclusively on life beyond Earth, while astrobiology also includes aspects related to the origin and evolution of life on Earth.

This temporal sequence of events that has led to the development of astrobiology as a young science worldwide, moving from the search for signals from technologically advanced civilizations to the biological perspective of seeking life, has also occurred in Argentina, a country that played a prominent role in the origins of this science in Latin America.

This article aims to summarize the most significant milestones that contributed to its development in the country, as well as to provide a brief state of the art at the time of writing this text.

4.2 The Beginnings of the Field: An Evolving Science

Just as with the exploration of planetary bodies in the universe, during the fifteenth century, the American continent represented a new world waiting to be discovered, thereby opening a door to the development of new knowledge. The first glimpses of cultural and scientific activity in South America are often attributed to religious orders, which conceived science in a manner closely aligned with modern perspectives that developed post-seventeenth century. As the historian of science José Babini recounts, "if at first no science is conducted in America, Europe conducts science with America." In particular, in the territory that is now Argentina, the Jesuit order predominated in carrying out the first cultural and educational activities. Babini refers to this as "the first rudiments of the sciences." Thus, the Jesuits, under the name "Society of Jesus," began to introduce natural sciences and astronomy, among other branches of knowledge (Babini, 1949). They also established the Colegio Máximo, which provided instruction in philosophy and theology, primarily aimed at members of their order, and which served as the foundation for the University of Córdoba, founded in 1622.

One notable figure who emerged from this institution was Father Buenaventura Suárez, a Jesuit born in the region that would later become Argentina, who made early regional developments in astronomy, starting in 1706 (it is worth noting that the first telescopes were developed in the seventeenth century). This coincided with paleontological discoveries, such as that of the Dominican friar Manuel Torres, and the work of naturalists, including the Jesuit Gaspar Xuárez in the field of botany, among others (Babini, 1949; Parodi, 1964).

However, the earliest known indications of interest in exploring the questions posed by astrobiology, at least known until the date, appeared in the region at the beginning of the nineteenth century. A key figure during this period was Manuel

Moreno (Buenos Aires, 1781–1852), a politician, military officer, and physician, brother of the Argentinian national hero Mariano Moreno. His notable positions included official in the Royal Treasury of the Viceroyalty of Rio de la Plata, sublieutenant in the defense of Buenos Aires against the British invasions in the region, and Chief Officer of the Secretariat of State in 1813. Moreno later emigrated to the United States, where he studied medicine at the University of Maryland, Baltimore. After receiving his bachelor's degree and doctorate in medicine, he returned to Buenos Aires in 1821, after Argentina became an independent country in 1816.

Between 1821 and 1823, several significant scientific institutions were founded, including the University of Buenos Aires (UBA), the National Academy of Medicine, the Literary Society, and the Society of Physical-Mathematical Sciences. The University of Buenos Aires not only operated as a university in contemporary terms but also functioned as a "ministry," overseeing all education, including high school education. It was divided into six departments: elementary education, preparatory studies (high school education), medicine, exact sciences, sacred sciences, and jurisprudence.

Manuel Moreno, who also presided over the National Academy of Medicine, was responsible for the chair of experimental chemistry in the Department of Preparatory Studies at UBA (De Asúa, 2011). The evidence linking Moreno to topics related to what we now understand as astrobiology pertains to a newspaper that existed at that time called "La Gazeta de Buenos Ayres," founded by his brother Mariano Moreno, which published official documents, resolutions, and governmental decrees, as well as current news from both abroad and locally. According to Guillermo Lemarchand in a chapter of the book "Astrobiology: From the Big Bang to Civilizations," Moreno discussed in La Gazeta the origin of life on Earth and the possible existence of life on other planets (Lemarchand, 2010a).

Lemarchand also highlights Vicente Fidel López (Buenos Aires, 1815–1903), a historian, lawyer, and politician, son of the composer of the Argentine National Anthem. López, who also served as rector of UBA, is said to have discussed the work of the German scientist and explorer Alexander von Humboldt, entitled "Kosmos," a five-volume treatise on science and nature. In the first volume, Von Humboldt describes the physical nature of outer space and Earth, and López, based on this book, speculated in a series of letters about volcanism and its potential to generate life on other planets (Lemarchand, 2010a).

In 1924, Argentina participated in a joint effort by several countries to detect artificial radio signals from Mars with the purpose of searching of advanced civilizations on that planet. This initiative was led by David Peck Todd, a US American astronomer who proposed that all radio stations on Earth should shut down during Mars's close passage, as the planet would experience a minimal distance from Earth that year. Following Todd's instructions, the Argentinian Navy executed successive cuts of 5 min/h in radio stations to listen for the presence of signals (Lemarchand, 2010a). This idea stemmed from the original concept posed by Nikola Tesla in 1899, which suggested the possibility that a civilization could be transmitting signals from Mars.

Thirty years after Vicente Fidel López's death, Carlos Varsavsky (1933–1983), an Argentinian physicist specialized in astrophysics, was born. During his life, Varsavsky emigrated to the United States after completing his high school education in Buenos Aires, where he obtained a master's degree and a bachelor's degree in Physics Engineering from the University of Colorado. He then earned his doctorate in Astronomy from Harvard University in 1959. His thesis work related to atomic transitions of astrophysical interest, remained a reference work for several decades.

In 1958, a year before Varsavsky received his doctorate, the Commission for Astrophysics and Radio Astronomy (CAR) was established in Argentina, following the installation of a solar interferometer at 86 MHz in the Facultad de Agronomía (Faculty of Agronomy) at UBA. The CAR was composed of Drs. Félix Cernuschi, Enrique Gaviola (president), and Engineer Humberto Ciancaglini, and arose from Dr. Merle Tuve's visit to Argentina, from the Carnegie Institution of Washington (CIW), USA. CIW aimed to establish the observatory known as the "Carnegie Southern Station for Radio Astronomy," due to the need for signals from southern latitudes. Several students, including Fernando Raúl Colomb (1939–2008) and Valentín Boriakoff (1938–1999), who later played crucial roles in the early search for intelligent extraterrestrial civilizations in Argentina, participated in the construction of the solar interferometer. Although the interferometer did not yield scientifically useful results, it fulfilled Dr. Tuve's initial objective of sparking interest in radio astronomy in Argentina (Historical Review of IAR, n.d.; Bajaja, n.d.).

Starting in 1959, radio telescopes gained additional prominence following a paper published in the journal *Nature*. The article, authored by two physicists, Giuseppe Cocconi and Phillip Morrison from Cornell University, USA, titled "Searching for Interstellar Communications," proposed that the hydrogen line at 21 cm (corresponding to neutral hydrogen, the most abundant element in the universe) detected by radio telescopes could be used for communication with extraterrestrial civilizations (Cocconi & Morrison, 1959). Following Cocconi and Morrison's publication, Frank Drake conducted the first search for these signals of "intelligent life" in the context of the "Ozma" Project in the 1960s. This project was developed at the National Radio Astronomy Observatory (NRAO) in Green Bank, West Virginia, USA.

Shortly after the CAR was created, in 1960, Carlos Varsavsky returned to Buenos Aires, where he joined the Astrophysics group in the Department of Physics at the Facultad de Ciencias Exactas y Naturales (Faculty of Exact and Natural Sciences) at UBA. He also served as a full professor of Physics at that university until 1966.

Due to the rise of radio astronomy in the country, an initiative involving several institutions such as the National Scientific and Technical Research Council (CONICET), the Provincial Commission for Scientific Research of Buenos Aires (CIC), the National University of La Plata (UNLP), and UBA, combined with negotiations between Dr. Tuve and Dr. Bernardo Houssay, then president of CONICET, led to the creation of the National Institute of Radio Astronomy (INRA) in 1962. The institute later adopted its current name, Argentinian Institute of Radio Astronomy (IAR).

Thus, in 1963, the construction of the 30 m diameter parabolic antenna, part of the IAR, commenced alongside civil works at the chosen location, very close to the city of La Plata, Buenos Aires province. The CIW from the USA also contributed parts of the first antenna. At that time, this became the largest radio telescope in South America. Among the functions of the IAR were to "promote and coordinate the research and technical development of radio astronomy and collaborate in education." Consequently, some scientists and engineers traveled abroad to enhance their knowledge and acquire experience in observing techniques for the 21 cm hydrogen line (Historical Review of IAR, n.d.).

On March 26, 1966, the IAR was officially inaugurated with the completion of the first antenna. Varsavsky became the first director of the Institute and played a foundational role.

As mentioned earlier, the establishment of the IAR occurred during a period when the proposal by Cocconi and Morrison regarding the potential detection of signals of intelligent life in the universe using radio astronomy tools was gaining significant attention, especially after Frank Drake initiated its first experimental search. Furthermore, the formulation of the now-famous "Drake Equation" by the scientist in 1961 provided a theoretical platform and a basis from which to establish a starting point for the exploration of the universe in the search for life.

In this unique period for radio astronomy, it is no coincidence that Varsavsky developed an interest in the subject and, notably, delivered a lecture in 1965 on the topic. A comment about his lecture appeared in the magazine Panorama:

> "Not long ago, around April 1965, a crowd of curious individuals flooded the old halls of the Faculty of Exact Sciences on Peru Street to listen to a young scientist proposing a spectacular topic: the possibility of life in the Universe. Many attendees expected a sort of master lecture, where the speaker would delight them with the most fantastic speculations and the most exciting assertions. But they were disappointed, precisely because the person standing at the lectern was a man of science: instead of indulging in speculation, he did what was both his habit and his profession, he broke a piece of chalk and began to perform calculations on the blackboard. Now, three long years after that lecture, radio astronomer and astrophysicist Carlos Varsavsky remains true to that tradition of modern science: he prefers calculation over unreflective theorizing. At 35 years old, he is preparing to once again impress both colleagues and the curious—those who have already been amazed by his widely recognized work in radio astronomy—with a study on the possibility of life in the Universe, a book that will likely bear that name ("Life in the Universe"), and which the Carlos Pérez publishing house is set to release in the coming weeks". (Panorama magazine, 1968)

This article from Panorama magazine also featured a brief review of Varsavsky's book "Vida en el Universo" ("Life in the Universe"), published in 1968. Regarding the topic of the search for intelligent life in the cosmos, this was the first book written by an Argentinian scientist, and perhaps the first by a Latin American scientist on the subject. The text is characterized by its excellent popular science communication style and is today considered one of the references in the field of public science communication.

In the introduction to the book, one can read the following paragraph:

"The anguish that a child feels when he first questions what exists beyond Earth is one of the problems humanity has had to overcome in contemplating the Universe. For a long time, humanity has sought to explore the mysteries of space, but only relatively recently has it approached these questions with a genuinely scientific perspective. A clear and serene night is an invitation to think about the meaning of what our eyes can see. In that moment: who has not stopped to ponder what lies beyond what our eyes can perceive? who has not felt moved at some point by the vastness of the spectacle before us? This sense of wonder that many have felt arises because for thousands of years, in the Western world, humanity considered itself the center of events, processes, situations, and worlds". (Varsavsky, 1968)

4.3   The SETI Project in Argentina

In Argentina, the IAR began to host its first students around the 1960s, including Fernando Colomb, who had participated in the construction of the solar interferometer by the CAR. He served as a technical assistant at the IAR from 1963 to 1965, the year he earned his degree in Physical Sciences from UBA.

By 1966, the year of the IAR's inauguration, Colomb traveled to the same radio astronomical observatory where Frank Drake conducted the first search for intelligent signals, the NRAO, where he worked as a research assistant.

Contemporary to Colomb, Valentín Boriakoff (1938–1999) worked at the IAR. He was an electromechanical engineer with a focus on electronics from UBA. He engaged in electronic engineering tasks before emigrating to the U.S. and the NRAO, just like Colomb. In 1973, Boriakoff earned his doctorate from Cornell University after working with Frank Drake on radio frequency observations of pulsars. Additionally, he served as an associate researcher at the National Astronomy and Ionospheric Center and other institutions, and worked as an adjunct professor in the Department of Astronomy at the University of Oklahoma and the Phillips Laboratory, a research laboratory of the U.S. Air Force, among others (Campbel et al., 2000).

In the field of astronomy, Boriakoff achieved notable accomplishments, including the detection of the first millisecond binary pulsar (Campbel et al., 2000). Regarding the search for life in the universe, he also made contributions related to the Voyager mission, particularly one linked to the famous "golden record." Frank Drake was one of the scientists involved in the task that included the mentioned record and recruited Boriakoff among others. There was a pressing need to figure out how to place images onto a metal disc in record time, and Drake entrusted this task to Boriakoff, who resolved the issue with a fundamental contribution (Scott, 2020). As Guillermo Lemarchand recounts, *"... in 1977, Boriakoff solved the method for recording images at a speed of 33 1/3 r.p.m. for the interstellar disc of the Voyager I and II spacecraft, thus contributing to the design of an interstellar message intended for a hypothetical extraterrestrial civilization."* Indeed, Carl Sagan mentions Boriakoff in his book *"The Sounds of Earth"* (translated in its Spanish version as *"Murmullos de la Tierra"*). As an additional note, a rather peculiar photograph of Boriakoff appears as one of the images included in the golden record of the Voyager spacecraft (Lemarchand, 2010a).

On another note, by the early 1980s, specifically in 1982, the International Astronomical Union (IAU) established Commission 51, titled *"Bioastronomy: Search for Extraterrestrial Life,"* which was later renamed simply *"Bioastronomy"* in 2006. However, this commission, still active (categorized as commission F3), currently exists under the name of "Astrobiology". Shortly after its creation, this commission organized a symposium in Boston, which was the first official IAU symposium dedicated to the study of the search for life in the universe (Lemarchand, 2010a). Félix Mirabel, a Uruguayan scientist naturalized as Argentine, who graduated with a degree and doctorate in astronomy from the National University of La Plata (Argentina) and a doctorate in philosophy from UBA, participated in this symposium. Mirabel, one of the most prominent astronomers in Argentina, proposed at this meeting the establishment of a project to search for signals of extraterrestrial origin in South America (Mirabel, 1984).

Subsequently, Mirabel worked on the search for interplanetary communication signals in the anti-maser hole of the formaldehyde molecule with the 140-foot radio telescope of the NRAO. He also took part in organizing meetings on this topic upon his return to Argentina in 1983 (personal communication with Dr. Mirabel).

In fact, Dr. Mirabel was one of the invited speakers at the *"First Interdisciplinary Meetings on Intelligent Life in the Universe,"* organized by a group of students from the Facultad de Ciencias Exactas y Naturales (Faculty of Exact and Natural Sciences) at UBA in 1985 (Lemarchand, 1986). This same group of students formed what was called the *"Astrophysics Commission,"* whose objective was to stimulate studies of physics applied to space, as there was no astronomy degree program at UBA. This commission also published a magazine called *"Astrophysics"* for several years, covering various topics, including those related to astrobiology (Fig. 4.1).

At the student group, the primary promoter and organizer of the Commission and the symposium was Guillermo Lemarchand, who at the time was a young physics student at UBA. The symposium, which lasted 3 days, was highly enriching. Approximately 500 participants attended, including students, the general public, and invited guests such as Félix Mirabel, Robert Bruce Cow from NASA, and Fernando Raúl Colomb, who had by then become the director of the IAR (Lemarchand, 1986; Historical Review of the IAR, n.d.). The symposium facilitated the approach and opening of the theme of the search for life in the universe within the country. Lemarchand, now physicist, recounts in an interview conducted in 2003: *"I was very fortunate because in the early years of my studies, I had the opportunity to organize, along with other students, the first Interdisciplinary Symposium on Intelligent Life in the Universe, which took place at the Faculty in 1985. This put me in contact with people working abroad, with whom we began to collaborate in Argentina on the SETI project"* (Olivella, 2003).

As part of the initiatives arising from the organization of this event, a cooperation agreement was signed between the IAR and The Planetary Society for the establishment of the SETI project in Argentina. This project was the first and only one in Latin America.

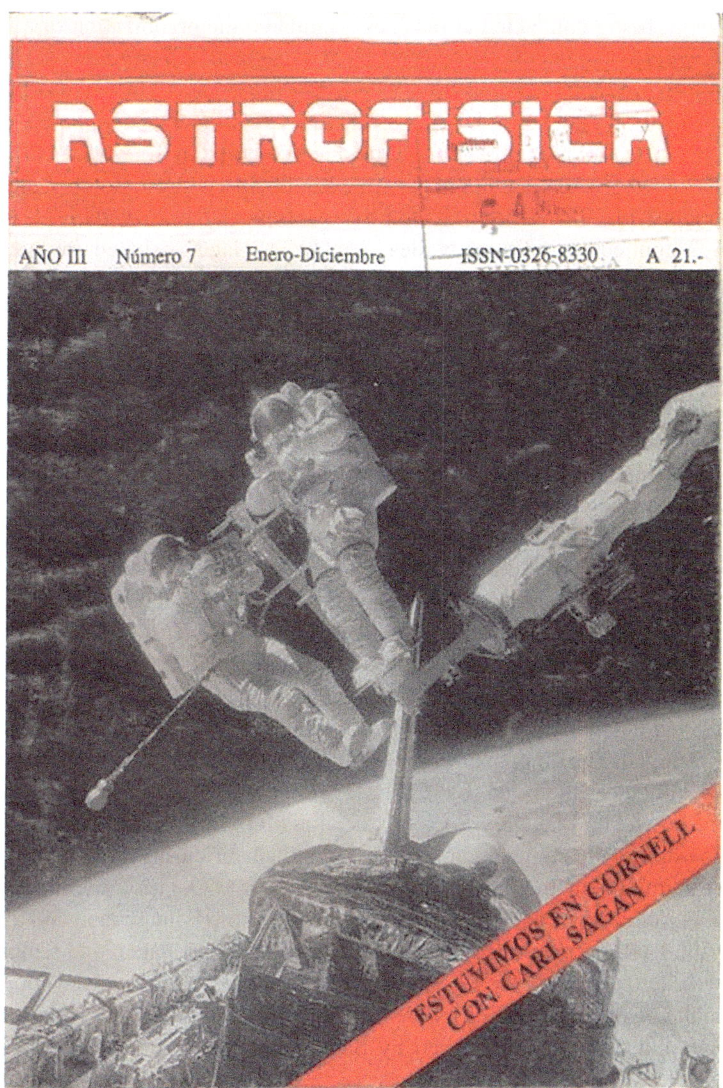

Fig. 4.1 An issue of the magazine "*Astrophysics*," produced by students of the Physics program at Facultad de Ciencias Exactas y Naturales, UBA, where topics related to astrobiology, among others, were discussed

On October 7, 1986, a milestone was achieved: the antennas of the IAR made their first observations of the sky in search of signals from civilizations that might originate from nearby stars. The research was conducted by Dr. Colomb, whose research group included Guillermo Lemarchand and María Cristina Martín. At the time of this search, only a few observations had been made from the Southern Hemisphere, specifically in Australia (Colomb et al., 1992).

The work being conducted at the IAR stimulated the organization of a seminar on the "*radioastronomical, social, and legal aspects of detecting electromagnetic signals of extraterrestrial origin*" led by the Argentine jurist Aldo Armando Cocca. The jurist, who studied law at UBA, was the first Argentinian to complete a thesis in Space Law (1953), which turned out to be one of the first in the world. For this reason, he is known as the "founder of the Argentinian Space Law." Notably, among his achievements, in 1954, Cocca created the concept of the "*Common Heritage of Mankind*" applied to celestial bodies, which was later included in the "*Moon Agreement*" of the United Nations (1979). Cocca was also the ideologue behind the Planetarium of Buenos Aires while serving as the Secretary of Culture and Social Action for the Municipality of Buenos Aires.

Focusing again on the work at the IAR, there was a SETI-related initiative in the United States carried out by The Planetary Society (founded in 1980 by Carl Sagan, Bruce Murray, and Luis Friedman), which developed the META project (an acronym for "*Megachannel Extraterrestrial Assay*") that has been operational since 1985, proposed and led by Professor Paul Horowitz, a physicist at Harvard University. The META project arose from the development of a portable spectral analyzer, referred to as "Suitcase SETI," which was put into use in 1983 alongside the Harvard/Smithsonian radiotelescope at the Oak Ridge Observatory in Harvard, Massachusetts (the project was originally called "*Sentinel*"). There was interest from the SETI community in the United States to extend the observation range of the META project to the Southern Hemisphere.

In 1988, Lemarchand visited Harvard University, where the SETI project was being developed. The objective was to establish the basis for an agreement with The Planetary Society, then led by Carl Sagan. This agreement was drafted by Lemarchand at The Planetary Society's headquarters in Pasadena, California. The venue for signing this agreement occurred months later during a conference on the search for intelligent life in the universe, held at the Ontario Science Centre in Canada, organized by The Planetary Society, which both Lemarchand and the president of the IAR, Fernando Colomb, attended. There, Carl Sagan and Colomb formalized the agreement for the signing of the convention, which involved cooperation between the National Council for Scientific and Technical Research (CONICET) of Argentina and the organization led by Sagan (for historical pictures and further details see Lemarchand, 2010a).

As Lemarchand himself recounts, "*the agreement involved the construction of a replica of a spectral analyzer with 8.4 million channels, featuring a spectral resolution of 0.05 Hertz per channel, which was operational at the 26-meter radio telescope of Harvard University.*" In this manner, with an antenna located in the northern hemisphere (the Harvard University radio telescope in OAK Ridge, USA) and another antenna in the southern hemisphere (belonging to one of the IAR radio telescopes), it was possible to conduct mapping in search of extraterrestrial signals, continuously observing the entire sky accessible from Earth (Lemarchand, 2010a).

To realize this project, two Argentinian engineers, Eduardo Horrell and Juan Carlos Olade, also participated, traveling to Harvard University with funding from The Planetary Society. There, they received training from Professor Paul Horowitz for the construction of a spectral analyzer that was installed at that institution. The project was named META II and was officially inaugurated in Argentina in 1990, thus becoming the "sibling" of the META project in the United States.

The IAR antenna conducted sky observations for this project for nearly 20 years (Lemarchand, 1994; Colomb et al., 1992). By 2003, Lemarchand noted that from the inauguration of META II until that time, approximately 20 billion distinct signals originating from outer space had been analyzed. Thus, Argentina became the only developing country with a SETI project.

In addition to his role as a pioneer of Astrobiology in the country, it is noteworthy that Lemarchand emphasized the promotion of scientific ethics and the idea of social responsibility that weighs upon scientists. As part of this motivation, in 1988, he organized, along with the Commission of Astrophysics he presided over, the *"First Latin American Symposium on Scientists, Peace, and Disarmament"* at the Facultad de Ciencias Exactas y Naturales (Faculty of Exact and Natural Sciences), UBA, Argentina. As a product of this meeting, Lemarchand created what is known as "The Buenos Aires Oath," an optional oath for scientists (similar to the Hippocratic oath) that began to be used from that year in the graduation ceremonies of that Faculty (Lemarchand, 2010b). The text focuses on the use of scientific knowledge in favor of ethics and peace:

> *"Being aware that science and, in particular, its applications can cause harm to society and humanity in the absence of adequate ethical controls, I firmly commit that my capacity as a scientist will never serve purposes that undermine human dignity, guided by my personal convictions, based on an authentic understanding of the situations around me and the possible consequences of the results derived from my work, not placing remuneration or prestige first, nor subordinating myself to the interests of employers or political leaders. The scientific research I conduct will be for the benefit of humanity and in favor of peace".*

It was a time when fears of nuclear war were increasing, particularly regarding what was termed "nuclear winter." Sagan's well-known statements on this matter, shared by Lemarchand, reflected a concern for the social aspects of science. These declarations have been recorded in interviews, articles, and his own books.

Motivated by common interests, in 1992, Sagan invited Lemarchand to the Center for Radio Physics at Cornell University. There, the Argentinian physicist worked as a Visiting Fellow alongside the American astronomer for one year.

Like Sagan, Lemarchand also engaged in science communication, addressing the search for intelligent life in the universe. Among his numerous articles and interviews, his book "El llamado de las Estrellas" (*"The Call of the Stars"*) (Lemarchand, 1992) stands out. It contains ten chapters, throughout which he describes the history of concepts of life beyond Earth, the origin and evolution of the universe, life and intelligence, the methodology used to detect exoplanets (which was at its begginings at the time), and among other topics, a discussion on the use of radio astronomy to detect signals from extraterrestrial civilizations and the development of SETI in Argentina.

After nearly 20 years of operation, the META II project required the renewal of its equipment. For this, several engineers needed to travel again to the United States to construct new analyzers, as had been done in the past. However, the project could not continue due to budgetary constraints and also because Guillermo Lemarchand, a fundamental pillar of the project in Argentina, became the Chief Consultant for the United Nations Educational, Scientific and Cultural Organization (UNESCO).

4.4 Stage of New Approaches in Astrobiology in Argentina

In the final years of the SETI project in Argentina, Lemarchand collaborated with two researchers: Dr. Andrea Buccino, who at the time held a degree in Physics and is now a researcher at CONICET, and Dr. Pablo Mauas, a physicist and also a CONICET researcher—both affiliated with the Instituto de Astronomía y Física del Espacio (Institute of Astronomy and Space Physics) (IAFE, UBA-CONICET). Lemarchand was not only a researcher at the IAR (Instituto Argentino de Radioastronomía) but also at the Center for Advanced Studies of the University of Buenos Aires.

Together, they developed a series of studies beginning around 2002, focusing on the effects of stellar radiation on planetary habitability—particularly ultraviolet (UV) radiation. These investigations emerged in the context of a growing international interest, as the first discoveries of extrasolar planets had recently been made: the first exoplanet in 1992 and the first exoplanet orbiting a Sun-like star in 1995. The research conducted in Argentina centered on habitability in exoplanets, specifically examining the impact of stellar flares emitted by so-called "red dwarfs" (dM-type stars).

Using theoretical approaches and modeling, the studies proposed a method for estimating the UV habitable zone in hypothetical extrasolar planets. These works considered the dual role of stellar UV radiation: on one hand, as a necessary factor for the origin of life—for instance, by acting as a catalyst in the formation of prebiotic organic molecules—and on the other, as a limiting factor, given that UV radiation can also be harmful to life (see, for example: Buccino et al., 2006, 2007).

Until that point, these studies had been conducted within a purely astrophysical theoretical framework. In 2007, having recently graduated with a degree in Biological Sciences from the Faculty of Exact and Natural Sciences at the University of Buenos Aires (FCEyN-UBA), a personal initiative led me to contact the group conducting this research. I believed that introducing an interdisciplinary perspective—one that incorporated biology and a biologist into the research team—would represent a substantial improvement in the approach to the subject, which had until then been addressed solely from an astrophysical standpoint.

The personal motivation to participate in this research topic stemmed from a longstanding interest in astronomy that preceded my formal academic training. This interest was academically reinforced through my participation in a public course offered at IAFE entitled *"Cosmología: una mirada a como pensamos que funciona el universo"* ("Cosmology: A Glimpse into How We Think the Universe Works"), taught by Dr. Gabriel Bengochea, physicist and now a researcher at

CONICET. This experience, in turn, led me to engage in amateur astronomy activities at the *"Club de Astronomía Ingeniero Félix Aguilar"* (*"Ingeniero Félix Aguilar Astronomy Club"*) (CAIFA), located in Vicente López, Buenos Aires Province, where I remained a member for several years. At CAIFA, some talks addressed the search for life in the universe, occasionally featuring special guests such as Guillermo Lemarchand.

As a result of the momentum generated by these events, and inspired by reading the publications on UV habitability, I sought to establish a potential interdisciplinary collaboration with astrophysicists. This led to the beginning of a personal research endeavor with Dr. Pablo Mauas at IAFE with the purpose of developing of a doctoral dissertation in astrobiology, in which we planned to conduct investigations that introduced a novel component to the study of UV habitability: the inclusion of biological experiments. During the first year, some preliminary experimental work was carried out in collaboration with Dr. Hugo Adamo of the Institute of Biological Chemistry and Physicochemistry (UBA-CONICET). These experimental efforts were later expanded through a collaboration I established with Dr. Eduardo Cortón from the Department of Biological Chemistry at FCEyN-UBA.

The doctoral research was interdisciplinary in nature, combining aspects of astrophysics—particularly stellar astrophysics—with biology, specifically microbiology. Its objective was to evaluate the effect of UV radiation on planetary habitability by using microorganisms as models of potential life forms on other planets. Based on this approach, the aim was to develop estimations to define a UV habitability zone in hypothetical extrasolar planets. Preliminary results were obtained by evaluating the dM-type star EV Lac (see e.g.: Abrevaya et al., 2009).

In addition, the research incorporated studies related to bioelectrochemistry, which were linked to the development and design of sensors for in situ life detection on other planets (Abrevaya et al., 2010). It is important to note that, up to that point, no prior studies of this nature had been conducted in Argentina under any of the aforementioned approaches, marking the beginning of laboratory-based experimental astrobiology research in the country.

In parallel with the development of the doctoral dissertation—and despite its divergence from the topic of the search for intelligent extraterrestrial life—in 2009, Guillermo Lemarchand organized the *"Second Ibero-American Postgraduate School of Astrobiology"* in Montevideo, Uruguay. The event featured the participation of researchers such as Dr. Gonzalo Tancredi from Uruguay, who co-organized the event; Dr. Antonio Lazcano, a pioneer in the study of the origin of life in Mexico; and Dr. Gustavo Porto de Mello, astronomer from Federal University of Rio de Janeiro (UFRJ), Brazil, among other researchers and students from various Ibero-American countries, myself included. As a culmination of the event, the book *"Astrobiología: del Big Bang a las civilizaciones"* (*"Astrobiology: From the Big Bang to Civilizations"*) was published (Lemarchand & Tancredi, 2010).

As part of this collaborative work with researchers from the region, my doctoral research also included a specific chapter involving scientists from Brazil. We initiated this joint effort with Dr. Iván Paulino-Lima, a biologist trained at the Federal University of Rio de Janeiro (UFRJ), whom I first met in 2009 during the

Astrobiology Graduate Student Conference held at the University of Washington in Seattle, USA.

In 2010, we carried out experiments with a team composed of Dr. Douglas Galante (astrophysicist), and Dr. Fabio Rodrigues (chemist) —both from the University of São Paulo (USP)—as well as Dr. Claudia Lage, a biologist from UFRJ. The Brazilian collaborators, who had prior experience in conducting inter-disciplinary studies under conditions simulating extraterrestrial environments, facil-itated the execution of a joint research project focused on lithopanspermia.

The experiments were conducted at the particle accelerator of the National Synchrotron Light Laboratory (LNLS, CNPEM) in Campinas, Brazil, and involved exposing extremophilic microorganisms to UV radiation in a vacuum, thereby experimentally simulating the conditions of low Earth orbit during an interplanetary journey. The results of this study led to a publication in the journal *Astrobiology*. This marked the beginning of a close collaborative relationship in the field of astro-biology between Argentina and Brazil (Abrevaya et al., 2011a, 2011b, 2023; Lage et al., 2012).

In March 2011, three and a half years after the beginning of my doctoral project, the PhD research was completed and the dissertation was submitted—marking the first doctoral thesis in astrobiology in Argentina. The thesis was entitled: "Effects of UV Radiation on Microorganisms for the Development of Habitability Models and the Search for Simple Life Forms on Solar and Extrasolar Planets" (Abrevaya, 2011).

Subsequently, during the postdoctoral stage, the foundations were laid for a col-laboration with astrophysicists and astronomers from the former Department of Geophysics, Astrophysics, and Meteorology, currently Departament of Astrophysics and Geophysics at the Institute of Physics, University of Graz, Austria. This col-laboration was carried out together with Prof. Dr. Arnold Hanslmeier and Drs. Martin Leitzinger and Petra Odert from the same institution. This interdisciplinary partnership also enabled the establishment of an international cooperation project, which I co-directed with Dr. Hanslmeier. The project, titled *"The Influence of UV Radiation from the Young Sun on Life"*, was supported and funded by the science ministries of Argentina and Austria (MINCyT and BMWF, respectively), for the period 2012–2015 (see e.g.: Abrevaya et al., 2013).

The effort to integrate interdisciplinary aspects of biology and astrophysics, developed during my doctoral and postdoctoral studies at IAFE (UBA - CONICET), was further expanded in Brazil through a second postdoctoral position under the supervision of Prof. Dr. Jorge Horvath, an Argentinian astrophysicist based in that country. The research was formalized at the Institute of Geophysics, Astronomy and Atmospheric Sciences of the University of São Paulo (USP) and the *"Núcleo de Pesquisa em Astrobiologia,"* with Dr. Horvath serving as the coordinator of the lat-ter. The project, titled *"Stellar Radiation and Its Influence on the Biosphere of Planetary Bodies"*, represented an extension and new approach to the studies previ-ously conducted and, once again, combined interdisciplinary astrobiological research involving both microbiology and astrophysics. This effort led to the con-solidation of two projects and a wide array of collaborations in the field of astrobiol-ogy, both in Brazil and internationally.

One of the projects that began to take shape during this period was entitled "*BioSun*" (Abrevaya et al., 2014), whose initial foundations were laid through the collaboration with researchers from the University of Graz, Austria, and further developed during the postdoctoral stay in Brazil with Prof. Horvath. Another line of research that also began to emerge at that time eventually evolved into the "*EXO-UV Program*", which continues to involve researchers from Austria, the United Kingdom, and Brazil, and currently Germany (Abrevaya et al., 2014).

The BioSun project is aimed at studying how radiation from our Sun in the past may have influenced the origin of life in the Solar System. Meanwhile, the EXO-UV program focuses on investigating how stellar radiation may affect the potential for life to exist on other planetary bodies.

During this postdoctoral period, I established a scientific collaboration with Dr. María Eugenia Varela, cosmochemist and geologist, a meteorite expert and researcher at CONICET, based at the "*Instituto de Ciencias Astronómicas, de la Tierra y el Espacio*" ("*Institute of Astronomical, Earth and Space Sciences*") (ICATE) in San Juan, Argentina. Given Dr. Varela's expertise in meteorites, her contributions enriched the interdisciplinary research conducted up to that point, including studies related to lithopanspermia within the framework of the BioSun project. This marked the beginning of a collaboration that continues to this day.

The consolidation phase of these ongoing lines of research began upon my return to Argentina in 2014, when I commenced a new stage as a CONICET researcher at the IAFE, following my postdoctoral work and a visiting research stay at the Harvard-Smithsonian Center for Astrophysics at Harvard University. Several collaborations were reactivated in the country. In addition to the ongoing projects on planetary habitability and lithopanspermia, research on biosensors for life detection—originally initiated during the doctoral studies—was resumed and continued until 2018. This new stage also involved the supervision of several students (García et al., 2014; Figueredo et al., 2015; Saavedra et al., 2018).

During this period, I also initiated a collaboration with Dr. Gerardo Juan M. Luna, an astronomer and CONICET researcher at IAFE, who provided valuable support for continuing the research endeavors undertaken upon my return to Argentina, thereby enabling the sustained development of astrobiology in the country. The studies on stellar radiation and planetary habitability were resumed with a renewed perspective, and a significant enhancement of the experimental approaches was achieved through the incorporation of Dr. Oscar Oppezzo as a collaborator. Dr. Oppezzo, a biochemist from UBA and an expert in photobiology and microbiology from the Radiobiology Department of the "*Comisión Nacional de Energía Atómica*" ("*National Atomic Energy Commission*") (CNEA), contributed substantially. Dr. Ana Forte Giacobone, a microbiologist and researcher within the same institution, also participated in the project for several years. Given Dr. Oppezzo's extensive experience in the biological effects of UV radiation, his involvement proved essential in advancing experimental studies on the impact of stellar flares on potential life forms in exoplanets, as part of the EXO-UV Program (see e.g.: Abrevaya et al., 2025).

In parallel, additional collaborative ties began to be established in the field of microbiology with Dr. Paula Tribelli, supported by Dr. Nancy López both CONICET researchers from the Department of Biological Chemistry at IQUIBICEN (UBA–CONICET). These collaborations focused on topics related to extreme halophilic microorganisms, hypersaline environments, and lithopanspermia—areas that had been jointly developed with Dr. Varela from ICATE.

After nearly a decade since the emergence of this new interdisciplinary strand of astrobiological research in Argentina—and following several years of joint work with Drs. Varela, Luna, Oppezzo, Forte-Giacobone, Tribelli, and López—the personal proposal to establish the *"Núcleo Argentino de Investigación en Astrobiología"* (*"Argentinian Research Unit in Astrobiology"*) "Astrobio.ar", www.astrobioargentina.org, began to take shape between 2015 and 2016. This initiative was brought to fruition with the support of the aforementioned scientists, who had been closely collaborating in recent years. The Hub, inspired by the efforts of colleagues in Brazil with their *"Núcleo de Pesquisas em Astrobiologia,"* was founded following a structure and guiding principles similar to those defined by the Faculty of Social Sciences at the University of Chile: *"Research Hubs are entities designed to strengthen collective research in various thematically relevant areas for the discipline, for society, and for inter- and intra-disciplinary work. The organization of a Hub can take different forms. For example (…) it may incorporate researchers from different institutions"* (FACSO, n.d.) (Fig. 4.2).

The *Núcleo Argentino de Investigación en Astrobiología*, Astrobio.ar, is currently composed of researchers from CONICET and CNEA. In addition, the Hub benefits from the collaboration of international scientists from countries such as Austria, Brazil, Colombia, Germany, the United Kingdom, and Spain, among others.

Thus, Astrobio.ar continues today as an extension of the astrobiology projects initiated over a decade ago, with the objective of conducting and promoting interdisciplinary research in the field of astrobiology in Argentina. This interdisciplinarity has emerged as a distinctive feature, setting these initiatives apart from previous astrobiology research in the country, and represents an innovative approach to contemporary astrobiology by integrating seemingly distant scientific

Fig. 4.2 Logo of the *"Núcleo Argentino de Investigación en Astrobiología"* (*"Argentinian Research Unit in Astrobiology"*) (© Lucila Gimenez and Ximena Abrevaya, 2015)

disciplines—such as astrophysics, microbiology, and geology. This approach is reflected in the research lines currently under development (e.g., Abrevaya & Thomas, 2017; Abrevaya et al., 2015, 2016, 2023, 2024a; Abrevaya, 2012a).

Among Núcleo's most notable achievements is the first experimental determination—conducted under laboratory conditions at the international level—of the impact of stellar flares (in the ultraviolet radiation range) on life, particularly on microorganisms (Abrevaya et al., 2020a). This study, conducted within the framework of the EXO-UV Program, focused on the closest known exoplanet to the Solar System at the time, Proxima b, which orbits the dM-type star Proxima Centauri. Further studies focused on the TRAPPIST-1 planetary system (Abrevaya et al., 2024b).

It is worth highlighting that many of the topics developed not only contribute to advancing knowledge within the field of astrobiology, but also open pathways toward other areas of basic science and generate applied outcomes in related scientific fields—for instance, in areas with practical applications (e.g., Abrevaya, 2012b; Oppezzo et al., 2024).

The research Hub also aims to promote public outreach and science communication related to astrobiology. To this end, its members regularly produce content through their own initiatives, including popular science articles, public lectures, and media engagement. The Hub actively serves the broader community by participating in interviews and contributing to mainstream media outlets on a regular basis.

As part of other notable initiatives in the field of astrobiology, in 2015, together with Dr. Julio Valdivia Silva, an astrobiologist from Peru, we began discussions on establishing a network aimed at connecting and strengthening astrobiology research across Latin America. This initiative was formalized in 2018 with the participation of PhD(c) David Tovar, PhD(c) Angélica Leal, and PhD(c) Jimena Sánchez, under the name Red Latinoamericana de Astrobiología (Latin American Astrobiology Network). Today, the Astrobio.ar Hub is a founding member of the network and plays an active role as the official node representing Argentina.

4.5 Recent Projects Related to SETI

After the interruption of the META II project—part of the SETI initiative in Argentina—several decades ago, a new effort has emerged which can also be situated within the domain of the search for extraterrestrial intelligent life. This initiative is the OTHER Project (acronym for *"Otros Mundos, Tierra, Humanidad y Espacio Remoto"*, meaning *"Other Worlds, Humanity and Remote Space"*), developed at the Universidad Católica de Córdoba (UCC).

The OTHER Project is led by Jesuit priest and astronomer José Funes, former director of the Vatican Observatory (2006–2015) and currently a CONICET researcher at the Universidad Católica de Córdoba. Over the course of his career, he has focused on the study of eclipsing binary stars, star formation in nearby galaxies, and black holes located at the centers of disk galaxies. More recently, he has expanded his research to include the scientific exploration of the possible existence of intelligent life in the universe.

OTHER, founded by José Funes upon his return to Argentina along with a group of collaborators, is presented as a "laboratory of ideas" that seeks to approach the search for inhabited worlds through the lenses of science, philosophy, and religion. Among its core aims is to address and reflect on questions such as the potential impact that the discovery of an extraterrestrial civilization might have on the philosophical, social, and religious conceptions of our own civilization (Funes et al., 2016). The project uses the Drake Equation as a conceptual framework, with a particular focus on its final two parameters, which relate to social factors (Funes et al., 2019). This interdisciplinary group includes several Argentinian scientists, primarily astronomers, such as Dr. José Funes (UCC) and Dr. Marcelo Lares (IATE), as well as scholars like Dr. Mariano Asla (Universidad Austral), Dr. Lucio Florio (Pontificia Universidad Católica Argentina), myself and professionals from a variety of other fields. OTHER hosts regular seminars dedicated to the discussion of these topics and is also engaged in scientific research, science communication, and educational initiatives.

The Argentinian Research Unit in Astrobiology, collaborates with the OTHER project on topics related to the scientific search for extraterrestrial intelligent life. As a result of this collaborative effort, a book was recently published—authored by Drs. Funes, Lares, Abrevaya, Asla, and Florio—that addresses the search for extraterrestrial intelligence from multiple disciplinary perspectives, including astronomy, biology, philosophy, and theology (Funes et al., 2023).

4.6 Concluding Remarks

In Argentina, the interest in the search for life in the universe began to take shape as a solid scientific endeavor with the launch of the SETI project in the late 1980s. This initiative was carried out in the country through the field of radio astronomy, led by astronomers and astrophysicists. It is worth noting that at that time, research linked to SETI was still in full swing, despite having begun several decades earlier. As a result of the investigations conducted within its territory, Argentina became the first and only country in Latin America—and in the Southern Hemisphere—to host an active SETI project.

Subsequently, at the international level, research in the field of exoplanets began to gain momentum following the first confirmed discovery of a planetary body outside the Solar System in 1992 and the first exoplanet orbiting a Sun-like star in 1995. In Argentina, the early 2000s marked the beginning of the first studies related to this topic within the astrobiological context, particularly focusing on stellar activity in dM-type stars and its implications for planetary habitability, conducted as part of stellar astrophysics research.

Around 2007, a new phase of investigation emerged in which astrobiology in Argentina acquired, for the first time, an explicitly interdisciplinary character with the incorporation of the biological sciences. The methodological approaches also evolved innovatively during this period, introducing laboratory-based experimental work. These efforts began within the framework of a doctoral thesis (2007–2011), which would become the first astrobiology dissertation in the country.

This interdisciplinary approach to research later expanded to include areas such as geology and cosmochemistry, among others, and gained institutional consolidation with the creation of the *Núcleo Argentino de Investigación en Astrobiología* (Argentinian Research Unit in Astrobiology) (Astrobio.ar). The Hub is currently expanding, incorporating new topics and promoting astrobiological research through both national and international collaborations. It operates alongside other recent initiatives, such as the OTHER project, which also contributes to the field's expansion and revisits some of the foundational aspects of astrobiology in Argentina initially linked to SETI.

Acknowledgements The author wishes to express her gratitude to Lic. E. Wolovelsky for the critical review of this manuscript, to Lic. G. Lemarchand for his valuable suggestions regarding the reviewed material, and to Dr. Félix Mirabel for his generous contribution of personal testimony.

Declarations

Competing Interests The authors declare no competing interests.

References

Abrevaya, X. C., Corton, E., & Mauas, P. J. D. (2009). UV habitability and dM stars: An approach for evaluation of biological survival Abstract. *Proceedings of the International Astronomical Union, 5*(S264), 443–445. https://doi.org/10.1017/S1743921309993073

Abrevaya, X. C. (2011). *Efectos de la radiación UV en microorganismos para la elaboración de modelos de habitabilidad y búsqueda de formas de vida simples en planetas solares y extrasolares.* Tesis doctoral para optar por el grado de Doctora en Ciencias Biológicas. Facultad de Ciencias Exactas y Naturales de la UBA.

Abrevaya, X. C. (2012a). Astrobiology in Argentina and the Study of Stellar Radiation on Life. *BAAA, 56,* 113–122.

Abrevaya, X. C. (2012b). Chapter 5. Features and applications of halophilic archaea. In O. V. Singh (Ed.), *Extremophiles: Sustainable Resources and Biotechnological Implications* (pp. 123–158). Wiley.

Abrevaya, X. C., & Thomas, B. C. (2017). Chapter 2. Radiation as a constraint for life in the universe. In P. H. Rampelotto, J. Seckbach, A. Sharov, & R. Gordon (Eds.), *Habitability of the universe before earth* (Astrobiology exploring life on earth and beyond) (pp. 27–46). Elsevier B. V.

Abrevaya, X. C., Mauas, P. J. D., & Cortón, E. (2010). Microbial fuel cells applied to the metabolically based detection of extraterrestrial life. *Astrobiology, 10,* 965–971.

Abrevaya, X. C., Cortón, E., & Mauas, P. J. D. (2011a). Flares and habitability. In *Proceedings IAU Symposium 286: Comparative magnetic minima: Characterizing quiet times in the Sun and stars* (Vol. S286, pp. 405–409). Cambridge University Press.

Abrevaya, X. C., Paulino-Lima, I. G., Galante, D., Rodrigues, F., Cortón, E., Mauas, P. J. D., & de Alencar Santos Lage, C. (2011b). Comparative survival analysis of Deinococcus Radiodurans and the haloarchaea Natrialba magadii and Haloferax volcanii, exposed to vacuum ultraviolet irradiation. *Astrobiology, 11*, 1034–1040.

Abrevaya, X. C., Hanslmeier, A., Leitzinger, M., Odert, P., Mauas, P. J. D., & Buccino, A. P. (2013). UV Radiation of the young Sun and its Implications for life in the Solar System. *Central European Astrophysical Bulletin, 37*, 649–654.

Abrevaya, X. C., Hanslmeier, A., Leitzinger, M., Odert, P., Horvath, J. E., Ribas, I., Galante, D., & Porto de Mello, G. F. (2014). The BIOSUN project: An astrobiological approach to study the origin of life. *Revista Mexicana de Astronomía y Astrofísica, 44*, 144–145.

Abrevaya, X. C., Galante, D., Nobrega, F., Tribelli, P., Rodrigues, F., Araujo, G., Gallo, T., Ribas, I., Sanz Forcada, J., Rodler, F., Porto de Mello, G. F., Leitzinger, M., Odert, P., Hanslmeier, A.,& Horvath, J. E. (2015). The use of synchrotron radiation in Astrobiology: Lithopanspermia studies and the Biosun project. In *25th RAU—Anual Users Meeting LNLS-CNPEM*.

Abrevaya, X. C. Caneiro A., Horvath J. E., Galante D., Wilberger, D. O., Vega-Castillo J., Rodrigues F., & Varela M. E. (2016). *Synthesis of halite under martian simulated conditions: A study with astrobiological implications.* LPSC Contribution No. 1903, p. 2134.

Abrevaya, X. C., Leitzinger, M., Oppezzo, O. J., Odert, P., Patel, M., Luna, G. J. M., Forte Giacobone, A. F., & Hanslmeier, A. (2020a). The UV surface habitability of Proxima b: First experiments revealing probable life survival to stellar flares. *Monthly Notices of the Royal Astronomical Society, 494*, L69–L74. https://doi.org/10.1093/mnrasl/slaa037

Abrevaya, X. C., Leitzinger, M., Oppezzo, O. J., Odert, P., Luna, G. J. M., Patel, M., Forte-Giacobone, A. F., & Hanslmeier, A. (2020b). Towards astrobiological experimental approaches to study planetary UV surface environments. *International Astronomical Union Proceedings Series, IAUS345 Origins: From the Protosun to the First Steps of Life, 14*(S345), 222–226. https://doi.org/10.1017/S1743921319002205

Abrevaya, X. C., Galante, D., Tribelli, P. M., Oppezzo, O. J., Nóbrega, F., Araujo, G. G., Ricardi, M., Rodrigues, F., Odert, P., Leitzinger, M., Varela, M. E., Sanz-Forcada, J., Gallo, T., Porto de Mello, G. F., Ribas, I., Rodler, F., Hanslmeier, A., & Horvath, J. E. (2023). Protective effects of halite to vacuum and vacuum—UV radiation: A potential scenario during a young Sun super-flare. *Astrobiology, 23*(3), 245–268. https://doi.org/10.1089/ast.2022.0016

Abrevaya, X.C., Kargl, G., Zivithal, S. , Varela, M.E., Caneiro, A., Vega-Castillo, J. Oppezzo, O.J., Wilberger, D, Galante, D., Rodrigues, F., Pereira Da Silva, E., Leitzinger, M., Odert, P. Hanslmeier, A. Tribelli, P.M., Lammer, H. Demergasso,, C., Horvath, J.E. (2024a) THE SISS (SALTS IN THE SOLAR SYSTEM) program: Studies on laboratory simulations with brines and NaCl crystals. EANA conference 2024. Available online: http://www.eana-net.eu/conferences/EANA2024/abstract_show.php?id=181

Abrevaya, X.C., Odert,, P., Oppezzo, O.J, Leitzinger, M., Luna, G.J.M., Guenther, E., Patel, M.R., Hanslmeier, A. (2024b) An experimental study of the biological impact of a superflare on the TRAPPIST-1 planets ABSTRACT Monthly Notices of the Royal Astronomical Society 535(2) 1616–1624 https://doi.org/10.1093/mnras/stae2433

Abrevaya, X.C, Odert, P., Leitzinger, M., Oppezzo, O., Luna, G. J. M., Patel, M.R., Hanslmeier, A. (2025) The EXO-UV program: Latest advances of experimental studies to investigate the biological impact of UV radiation on exoplanets *Solar System Research 59*(6) https://doi.org/10.1134/S0038094624602019

Babini, J. (1949). Historia de la ciencia Argentina (Vol. 46). Editorial Fondo de cultura económica, México. 218 pp.

Bajaja, E. (n.d.). *La Radioastronomía en Argentina.* Retrieved April 25, 2025, from http://tux.iar.unlp.edu.ar/boletin/art-bajaja.htm

Buccino, A. P., Lemarchand, G. A., & Mauas, P. J. D. (2006). UV radiation constraints around the circumstellar habitable zone. *Icarus, 183*, 491–503. https://doi.org/10.1016/j.icarus.2006.03.007

Buccino, A. P., Lemarchand, G. A., & Mauas, P. J. D. (2007). UV habitable zones around M stars. *Icarus, 192*, 582–587. https://doi.org/10.1016/j.icarus.2007.08.012

Campbell, D., Carlson, H., Muslimov, A., Noori, M., & Van Horn, H. M. (2000). Obituary: Valentin Boriakoff, 1938-1999. *In Bulletin of the American Astronomical Society, 32*(4), 1656–1657.

Cocconi, G., & Morrison, P. (1959). Searching for interstellar communications. *Nature, 184,* 844–846. https://doi.org/10.4159/harvard.9780674366688.c9

Colomb, F. R., Martín, M. C., & Lemarchand, G. A. (1992). SETI observational program in Argentina. *Acta Astronautica, 26,* 211–212. https://doi.org/10.1016/0094-5765(92)90098-4

De Asúa, M. (2011). *Una gloria silenciosa, dos siglos de ciencia en Argentina.* Editorial Libros del Zorzal. 316 pp.

FACSO. (n.d.). Retrieved from https://facso.uchile.cl/sociologia/archivo-departamento-de-sociologia/nucleos-de-investigacion.

Figueredo, F., Cortón, E., & Abrevaya, X. C. (2015). In situ search for extraterrestrial life: A microbial fuel cell-based sensor for the detection of photosynthetic metabolism. *Astrobiology, 15,* 717–727. https://doi.org/10.1089/ast.2015.1288

Funes, J., Lares, M., De los Rios, M., Martiarena, M., & Ahumada, A. (2016). OTHER: A multi-disciplinary approach to the search for other inhabited worlds. *BAAA, 59,* 1–3.

Funes, J. G., Florio, L., Lares, M., & Asla, M. (2019). Searching for spiritual signatures in SETI research. *Theology and Science, 17*(3), 373–381.

Funes, J. G., et al. (2023). *La búsqueda de vida extraterrestre inteligente: Un enfoque interdisciplinario.* Editorial Universidad Católica de Córdoba.

García, M., Abrevaya, X. C., & Gómez, M. N. (2014). Condiciones físicas de exoplanetas y microorganismos que habitan ambientes extremos. *BAAA, 56,* 407–410.

Hays, L. et al. (Eds.) (2015). *NASA astrobiology strategy.* Retrieved April 25, 2025, from https://astrobiology.nasa.gov/nai/media/medialibrary/2015/10/NASA_Astrobiology_Strategy_2015_151008.pdf.

Historical review of IAR (n.d.). http://www.iar.unlp.edu.ar/historia.htm

Lage, C., Dalmaso, G., Texeira, L., Bendia, A., Paulino Lima, I., Galante, D., Janot-Pacheco, E., Abrevaya, X. C., Azua-Bustos, A., Pellizari, V., & Rosado, A. (2012). Probing the limits of extremophilic life in extraterrestrial environment simulated experiments. *International Journal of Astrobiology, 11,* 251–256. https://doi.org/10.1017/S1473550412000316

Lemarchand, G. A. (1986). Jornadas Interdisciplinarias sobre Vida Inteligente en el Universo. *Astrofísica, 1,* 3–8.

Lemarchand, G. A. (1992). *El llamado de las estrellas.* Colección Lugar Científico.

Lemarchand, G. A., & Tancredi, G. (2010). Astrobiología: del Big Bang a las civilizaciones. Montevideo: UNESCO. 347 pp.

Lemarchand, G. A. (2010a). Una breve historia social de la Astrobiología en Latinoamérica. In G. Lemarchand & G. Tancredi (Eds.), *Astrobiología: Del Big Bang a las Civilizaciones* (pp. 24–52). Editorial de la Unesco, 347 pp.

Lemarchand, G. (2010b). Science for Peace in the Benefit of Humankind. *The Hippocratic Oath for Scientists concept.*

Lemarchand, G. A. (1994). Passive and active SETI strategies using the synchronization of SN1987A. *Astrophysics and Space Science, 214*(1), 209–223.

Lemarchand, G. A., & Tancredi, G. (2010). *Astrobiología: del Big Bang a las civilizaciones* (p. 347). UNESCO.

Mirabel, I. F. (1984). Búsqueda de vida extraterrestre: Desarrollos recientes y nuevas perspectivas. *Revista Astronomica Organo de la Asociacion Argengina Amigos de la Astronomia Buenos Aires, 230,* 2–4.

Olivella, P. (2003). *Entrevista a Guillermo Lemarchand.* Buscando vida entre las estrellas. Available online: http://www.fcen.uba.ar/prensa/noticias/2003/opinion_07abr_2003.html

Oppezzo, O. J., Abrevaya, X. C., & Giacobone, A. F. (2024). An alternative interpretation for tailing in survival curves for bacteria exposed to germicidal radiation. *Photochemistry and Photobiology, 100*(1), 129–136. https://doi.org/10.1111/php.13808

Panorama semanal (1968), "Vida en el universo", *Año vi n° 87, Buenos Aires.*

Parodi, L. R. (1964). Biografía. Gaspar Xuarez, primer botánico argentino. *Darwin, 13,* 195–208.

Saavedra, A., Federico, F., Cortón, E., & Abrevaya, X. C. (2018). An electrochemical sensing approach for scouting microbial chemolithotrophic metabolisms. *Bioelectrochemistry, 123,* 125–136. https://doi.org/10.1016/j.bioelechem.2018.04.020

Scott, J. (2020). *The Vinyl Frontier. The story of NASA's interstellar mixtape.* Bloomsbury Publishing.
Varsavsky, C. M. (1968). *Vida en el Universo, Buenos Aires, Carlos Pérez Editor.* 123 pp.
Wright, J. T. (2018). SETI is part of astrobiology. *arXiv preprint arXiv:1801.04868.* https://doi.org/10.48550/arXiv.1801.04868

Chapter 5
Astrobiology and Planetary Sciences in Brazil

Paulo Leme and Jorge Horvath

Abstract This chapter provides a comprehensive overview of the development of astrobiology and planetary sciences in Brazil, tracing its historical roots, key researchers, and institutional milestones. It highlights pioneering contributions in prebiotic chemistry, microbial ecology, planetary astronomy, and the simulation of extraterrestrial environments. The chapter presents landmark initiatives such as the foundation of AstroLab, the establishment of the Núcleo de Pesquisa em Astrobiologia (NAP-Astrobio), and the creation of the Sociedade Brasileira de Astrobiologia (SBAstrobio). Emphasis is placed on the interdisciplinary nature of astrobiology in Brazil, encompassing collaborations among physicists, chemists, biologists, geologists, and astronomers. Education and outreach programs, particularly the development of undergraduate and graduate courses, workshops, and public engagement, are also explored. Despite financial and political challenges, Brazilian astrobiology has grown into a vibrant and collaborative field, contributing significantly to international efforts in the search for life beyond Earth.

5.1 Pioneers of Astrobiology in Brazil

The study of the origin and evolution of life in the universe has a significant historical background in the global context. The first text referring to astrobiology published in Brazil was the book Introdução à Astrobiologia, written by Flavio Augusto Pereira (1958). This work compiled the available knowledge on the subject at the time and played an important role in disseminating the topic within the country. It included several fundamental concepts of astrobiology that remain relevant today,

P. Leme
Universidade Cruzeiro do Sul—retirado, São Paulo, Brazil

J. Horvath (✉)
Instituto de Astronomia, Geofísica e Ciências Atmosféricas, Universidade de São Paulo, São Paulo, Brazil
e-mail: foton@astro.iag.usp.br

D. Tovar, M. A. Leal (eds.), *Astrobiology and Planetary Sciences in Latin America*, https://doi.org/10.1007/978-3-032-01450-4_5

such as the concept of the solar ecosphere—now referred to as the habitable zone—as well as scientific discussions concerning the origin of life, the Urey–Miller experiment (Miller, 1953), evolutionary theory, and some of the potential life forms capable of surviving in extreme environments, such as lichens.

At the Department of Chemistry of the Federal University of Pernambuco (UFPE), Ricardo C. Ferreira played an undeniably important role in initiating studies related to biogenesis and the origin of life. In the 1980s, he began working with biological systems and was also a pioneer in molecular biology in Brazil. During the same decade, in collaboration with Constantino Tsallis (Brazilian Center for Physics Research—CBPF), he developed an interest in biogenesis and self-replication in biological systems (Tsallis & Ferreira, 1983; Ferreira & Tsallis, 1985), publishing several studies on the origin of life in the following years (Ferreira & Coutinho, 1993; Lins et al., 1996; Ferreira & Cavalcanti, 1997; Soares et al., 1997; Cavalcanti & Ferreira, 2001; Cavalcanti et al., 2004). Recipient of numerous national awards, Ferreira served on the advisory boards of many scientific organizations and held visiting professorships at several international universities. More than a pioneering researcher in topics related to astrobiology, he contributed significantly to the advancement of science in Brazil, particularly in the field of chemistry (Tsallis, 2008).

In the field of astronomy, Sylvio Ferraz-Mello (Professor Emeritus at the Institute of Astronomy, Geophysics and Atmospheric Sciences of the University of São Paulo—IAG/USP) has played a pivotal role since 1988 in advancing research in planetary dynamics and exoplanets (Ferraz-Mello, 1988; Ferraz-Mello et al., 1993; Nesvorny & Ferraz-Mello, 1997; Michtchenko et al., 2002; Léger et al., 2009). His work on the motion of Jupiter's Galilean satellites (Ferraz-Mello, 1964, 1966) led to a PhD awarded by the Université de Paris (France) in 1966. Ferraz-Mello has published nearly 100 articles in major scientific journals and has supervised numerous master's and doctoral theses in astronomy, contributing significantly to the consolidation of Brazilian astronomy. In recognition of his many important contributions, the International Astronomical Union (IAU) named asteroid 1983XF (5201) in his honor.

As a means of understanding prebiotic chemistry, Fernando Barros (Federal University of Rio de Janeiro—UFRJ) conducted research on the interaction between inorganic surfaces and biologically important molecules (Tessis et al., 1995; Vieyra et al., 1995). Dimas A. M. Zaia (State University of Londrina—UEL) also contributed to studies on prebiotic chemistry and the conditions required for the formation of peptide polymers on the early Earth (Zaia et al., 2002, 2008; Zaia, 2003, 2004; Carneiro et al., 2011). Carol and Kenneth Collins (State University of Campinas, São Paulo—UNICAMP) worked on analytical chemistry and the role of radiation in prebiotic chemical processes (Albarran et al., 1993; Collins et al., 2000).

The historically significant approach to understanding how concepts and the search for extraterrestrial life developed during the twentieth century, as explored by historian Eduardo Dorneles Barceló, began with his master's studies in 1991. His research paid particular attention to NASA's exobiology program and the Search for Extraterrestrial Intelligence (SETI) project. Later, as a result of his doctoral dissertation, he published the book Telegramas a Marte—la Búsqueda Científica de la Vida Extraterrestre y la Inteligencia (Rodrigues et al., 2012). In parallel, Jorge A. Quillfeldt (Department of Biophysics—Federal University of Rio Grande do Sul—UFRGS), an enthusiast of exobiology and astrobiology, launched a public outreach course in 1982, which would later become one of the first undergraduate courses in astrobiology in 2002 (Rodrigues et al., 2012).

5.2 CoRoT Committee—Brazil

As part of subsequent activities aimed at studying issues related to astrobiology, the CoRoT mission—Convection, Rotation et Transits planétaires—stands out. This was the first mission capable of detecting rocky exoplanets several times larger than Earth orbiting nearby stars, and was launched in 2006. The mission was led by the French Space Agency (CNES), in conjunction with the European Space Agency (ESA) and other international partners (Rodrigues et al., 2012). Brazil played an active role in the detection and physical study of exoplanets through a team led by Eduardo Janot Pacheco (IAG/USP), which conducted photometric, spectroscopic, and/or polarimetric measurements of planetary transits from Earth, including observations from Brazil and Chile. These efforts contributed to the discovery and confirmation of new planets, as well as mass measurements. In addition to scientific production, the project resulted in five master's theses and four doctoral dissertations. Team members included A. D. de Souza Jr., Amâncio Friaça, Sylvio F. Mello, Tatiana Michtchenko, Paulo Penteado, Douglas Galante, M. Emilio, J. R. de Medeiros, L. B. P. Andrade, A. Baglin, Silvia P. Alencar, Bruno C. Martins, Daniel de Freitas, and Luander Bernardes (Rodrigues et al., 2012).

5.3 First Brazilian Workshop on Astrobiology—I BWA

At the beginning of the twenty-first century, many researchers from different fields across Brazil were conducting isolated individual studies. Although most of them had not previously engaged directly with astrobiology in their academic careers, they can now be considered pioneers in this field in Brazil. It was during this time that they began exploring astrobiology as a multidisciplinary area of study. There was a clear demand within the scientific community to foster exchanges of experience and to promote interaction among professional researchers and graduate students from various disciplines who were either already working on or interested in initiating research in the diverse topics related to astrobiology. In response to this need, the First Brazilian Workshop on Astrobiology was held from March 20–21, 2006, in the city of Rio de Janeiro. This event was of fundamental importance for the development of the field in the years that followed. The meeting was organized by Amâncio Friaça (Institute of Astronomy, Geophysics and Atmospheric Sciences,

University of São Paulo—IAG/USP), Carlos Alexandre Wuensche (Astrophysics Division, National Institute for Space Research—INPE), Claudia Lage (Department of Biophysics, Federal University of Rio de Janeiro—UFRJ), Gustavo F. Porto de Mello (Valongo Observatory, Federal University of Rio de Janeiro—UFRJ), and Vivian H. Pellizari (Oceanographic Institute—University of São Paulo) (Rodrigues et al., 2012).

Other pioneering researchers formed part of the Scientific Committee: Adriana Valio (Center for Radio Astronomy and Astrophysics, Mackenzie University—CRAAM), Cezar Sá (Institute of Biological Sciences, University of Brasília—UNB), Eduardo Janot Pacheco (Institute of Astronomy, Geophysics and Atmospheric Sciences, University of São Paulo—IAG/USP), Jaime Fernando V. da Rocha (Institute of Physics, State University of Rio de Janeiro—UERJ), Jorge E. Horvath (Institute of Astronomy, Geophysics and Atmospheric Sciences—IAG/USP), and Waldenor Cruz (Institute of Biological Sciences, University of Brasília—UNB). The workshop hosted 104 participants from 23 different institutions across the country, as well as three participants from other Latin American countries, receiving a total of 60 abstract submissions. David Catling (University of Bristol, UK) and Janet Siefert (Rice University, USA) were invited international keynote speakers. Astronomy, biology, physics, chemistry, geology, and planetary sciences were among the fields represented, and the workshop provided the first collective platform in Brazil for discussing advancements and challenges in this emerging discipline, enabling interested participants to connect, exchange ideas, and explore potential collaborative projects (Rodrigues et al., 2012).

5.4 Other Pioneers of Astrobiology in Brazil

Researchers from various universities across Brazil have carried out extensive work using synchrotron light and other radiation sources to understand chemical reactions in the interstellar medium that may be significant for the formation of molecules (Ferreira-Rodrigues et al., 2011; Lago et al., 2004; Boechat-Roberty et al., 2005, 2009; Pilling et al., 2006a, 2006b, 2007, 2009; Neves et al., 2007; Andrade et al., 2009). Among other researchers working in the field of astrochemistry are Arnaldo Naves de Brito (University of Brasília—UNB and University of Campinas—UNICAMP), Boechat-Roberty (Federal University of Rio de Janeiro—UFRJ), Sergio Pilling (University of Vale do Paraíba—UNIVAP), Diana P. P. Andrade (University of Vale do Paraíba—UNIVAP), Manoel G. P. Man (Federal University of Santa Catarina—UFSC), and Enio Frota da Silveira (Pontifical Catholic University of Rio de Janeiro—PUC-Rio).

In the field of astronomy, scientific research on planetary habitability has been conducted by Gustavo Porto de Mello (Federal University of Rio de Janeiro—UFRJ) and collaborators (Michtchenko & de Mello, 2009). The chemical evolution of the galaxy has been extensively studied by Walter Maciel, Helio Rocha-Pinto, Amâncio Friaça, Roberto Costa, and others (Friaça & Terlevich, 1998; Rocha-Pinto et al., 2004; Maciel et al., 2003). Furthermore, the evolution of primitive bodies (asteroids and comets), including their chemistry, has been investigated by Daniella Lazaro, Enos Picazzio, and Amaury Almeida (Duffard et al., 2004; Picazzio et al., 2007).

Spectroscopic techniques—particularly Raman spectroscopy—have been employed to characterize life in extreme environments, pigments, and biosignatures by Luiz Fernando Cappa de Oliveira (Federal University of Juiz de Fora—UFJF), in collaboration with Howell G.M. Edwards from the University of Bradford in the United Kingdom (Edwards et al., 2002, 2004a, 2004b; de Oliveira et al., 2010).

The application of computational models to analyze evolution in a broad sense has been explored by José Fernando Fontanari (University of São Paulo, São Carlos campus), with a particular focus on key problems related to the origin and evolution of life (Alves & Fontanari, 1997; Alves et al., 2000; Silvestre & Fontanari, 2005; Fontanari et al., 2006).

5.5 Brazilian Antarctic Program—PROANTAR

The Instituto Oceanográfico da Universidade de São Paulo has participated in the Programa Antártico Brasileiro (PROANTAR) through a research initiative led by Dr. Vivian H. Pellizari. Her project focused on investigating microbial diversity across various cold environments, including marine water, marine sediments, lacustrine waters and sediments, mineral soils, permafrost, and glacial ice (Kuhn et al., 2009; Rodrigues et al., 2009; Teixeira et al., 2010). These efforts evolved into a new PROANTAR-funded project launched in 2009, which marked a significant shift by explicitly framing the Antarctic environment as an analogue for astrobiological studies—representing the first research initiative of its kind in Brazil.

As an outcome of the First Brazilian Workshop on Astrobiology (first BWA), a dedicated research group in astrobiology was established by Eduardo Janot Pacheco (Institute of Astronomy, Geophysics and Atmospheric Sciences—IAG/USP) and Claudia Lage (Institute of Biology—IB/UFRJ). This collaboration laid the groundwork for the creation of a multidisciplinary virtual research network named Astrobio-Brazil, which aimed to foster interdisciplinary cooperation among researchers from diverse scientific domains with an interest in astrobiology.

As a natural outcome of these initiatives, several doctoral dissertations were completed in fields related to astrobiology, astrochemistry, and prebiotic chemistry. Specifically, in astrobiology, the following PhD theses were presented: Douglas Galante (Laboratório Nacional de Luz Síncrotron—LNLS), under the supervision of Jorge E. Horvath (IAG/USP), focused on the biological implications of high-energy astrophysical phenomena (Galante, 2009); Ivan Paulino-Lima, supervised by Claudia Lage (UFRJ), conducted experimental simulations of extraterrestrial environments (Paulino-Lima, 2010); and Rubens Duarte, under the guidance of Vivian Pellizari (USP), explored microbial ecology in polar environments (Duarte, 2010).

5.6   Astrobiology Laboratory—AstroLab

The first Brazilian multidisciplinary and open-access laboratory fully dedicated to astrobiology, named AstroLab (www.astrobiobrazil.org), was established in 2009 by the Instituto de Astronomia, Geofísica e Ciências Atmosféricas (IAG/USP). It was coordinated by Eduardo Janot Pacheco (IAG/USP), with the collaboration of Ramachrisna Teixeira, director of the Abrahão de Moraes Observatory (IAG/USP), and Professor Tércio Ambrizzi (IAG/USP). One of the laboratory's primary research foci has been the environmental microbiology of Brazilian sites, the Atacama region, and Antarctica, with particular emphasis on the isolation of new species of extremophilic microorganisms. Both experimental and theoretical simulations have been conducted—experimental simulations employed specialized chambers and facilities to recreate environmental conditions found in outer space or on other planetary bodies, while theoretical simulations were carried out through numerical and computational modeling.

Several Brazilian funding agencies—such as the Fundação de Amparo à Pesquisa do Estado de São Paulo (FAPESP), the Conselho Nacional de Desenvolvimento Científico e Tecnológico (CNPq) through its Brazilian Antarctic Program (PROANTAR) and the Instituto Nacional de Pesquisas Espaciais (INEspaço), as well as the Coordenação de Aperfeiçoamento de Pessoal de Nível Superior (CAPES)—provided financial support for a wide array of research projects at AstroLab over nearly a decade. Postdoctoral fellowships were awarded to Douglas Galante (IAG/USP) for the development of a simulation chamber and to serve as the laboratory's operational coordinator; to Fabio Rodrigues (Institute of Chemistry—USP) to lead research in metagenomics and spectroscopic biosignatures, ultimately becoming the laboratory's chemistry coordinator; and to Rubens Duarte (Oceanographic Institute—USP), to study the microbial ecology of Antarctic ecosystems, taking on the role of biology coordinator at AstroLab.

The project, coordinated by Vivian H. Pellizari (Oceanographic Institute, University of São Paulo—USP), titled "The Antarctic Microbial Environment as a Model for Astrobiology Studies," was officially the first initiative at AstroLab in which astrobiology was the central objective. The project's goals were structured into three main components: (1) the isolation and identification of extremophilic bacteria from Antarctic samples; (2) the detection and characterization of cold-adapted and UV-resistant strains; and (3) the exposure of selected extremophilic strains to simulated extraterrestrial environments. The core research team included Claudia Lage (UFRJ) and Eduardo Janot-Pacheco (IAG/USP), along with their doctoral and postdoctoral students. International collaborations were established with David Gilichinsky (Institute of Physicochemical and Biological Problems in Soil Science, Russian Academy of Sciences), Dr. Armando Azúa-Bustos (Pontificia Universidad Católica de Chile), and Dr. Ximena Abrevaya (University of Buenos Aires, Argentina).

The laboratory was conceived as an open-access facility, designed to serve as a physical hub for interaction among both Brazilian and international astrobiology communities. All of its facilities are available to researchers interested in conducting projects related to astrobiology. These include simulation chambers, sample preparation laboratories, and a fully equipped infrastructure to support multidisciplinary scientific investigations.

5.7   Research Center in Astrobiology—NAP-Astrobio

Created with the purpose of providing institutional support to AstroLab and fostering a multidisciplinary environment where researchers from various institutes and departments at the University of São Paulo (USP), as well as from other universities, can interact, the Núcleo de Pesquisa em Astrobiologia—NAP-Astrobio (http://www.astrobiobrasil.org) is led by Jorge Ernesto Horvath (IAG/USP) as Scientific Coordinator and Eduardo Janot Pacheco (IAG/USP) as Deputy Scientific Coordinator. NAP-Astrobio consists of a Deliberative Council composed of four faculty members from different USP departments, along with 48 PhD-level researchers and 23 graduate students (Master's and PhD), encompassing the fields of astronomy, biology, physics, chemistry, geology, biomedical sciences, oceanography, and engineering.

The projects and activities currently underway within the NAP-Astrobio initiative are distributed across several key areas:

- **Astronomical Observations and Theoretical Studies:** While some researchers focus on the observation and characterization of exoplanets, including the identification of habitable zones within the Galaxy, others approach the problem from a theoretical standpoint, aiming to understand the environmental conditions present on planetary surfaces—particularly Mars and early Earth. A strong emphasis is placed on the effects of radiation. Additionally, the interface between the physical sciences and biology is actively explored, especially in relation to the thermodynamic description of evolutionary processes within the framework of irreversible thermodynamics.
- **Simulations of Space and Planetary Environments:** As noted earlier, the experimental simulation of extraterrestrial and planetary conditions constitutes a central component of AstroLab's work. These simulations enable researchers to assess the survivability of microorganisms and biological molecules under conditions analogous to those found beyond Earth. This line of research is essential to support life detection missions, whether through in situ spectroscopic techniques (e.g., onboard space probes) or through remote astronomical observations. In this regard, the availability of environmental simulation chambers is a critical asset, fostering the advancement of early-stage investigations in this field.
- **Microbiology of Extreme Environments:** The group has conducted extensive research on a variety of extreme environments, particularly those with the potential to serve as terrestrial analogs of extraterrestrial settings. Notable among these are cryoenvironments in Antarctica, high-altitude sites in the Andes, the

hyper-arid Atacama Desert, and the deep ocean floor. Studies have encompassed microbial ecology, biodiversity assessments, and the isolation and characterization of extremophiles.

- **Paleobiology:** Paleobiology enables the study of early life on Earth to inform our understanding of biological evolution and to support the detection of biosignatures—particularly through spectroscopic techniques—in planetary environments such as the Martian surface. Active research in this domain includes the application of Raman spectroscopy and collaboration with international laboratories.
- **Prebiotic Chemistry:** Research in prebiotic chemistry has been conducted at AstroLab, with a primary focus on Raman spectroscopic methods. A central objective of this line of inquiry is to investigate molecular complexity as it arises from natural catalytic substrates, such as clays.
- **Education and Public Outreach:** The group has played a prominent role in science education and outreach, organizing schools, workshops, and the production of educational materials. This work is especially important given the nascent status of astrobiology in Brazil and the scarcity of educational resources in Portuguese—a gap that can lead to misinformation or the dissemination of scientifically questionable content. The establishment of the Brazilian Astrobiology Network (Rede Brasileira de Astrobiologia) and the development of international partnerships have contributed significantly to mitigating these challenges.

The NAP-Astrobio has played a pivotal role in consolidating the field of astrobiology in Brazil through the implementation of several key initiatives, as outlined below:

(a) Participation in the establishment of the AstroLab laboratory in 2011, a facility dedicated to research and the training of undergraduate and graduate students in astrobiology.

(b) Organization of the international workshop São Paulo Advanced School of Astrobiology (SPASA) in 2011 (http://www.astro.iag.usp.br/spasa2011/home. html), which brought together 12 international speakers and 18 Brazilian researchers. This event engaged 120 participants and covered virtually all major topics in astrobiology, fostering academic exchange that led to numerous subsequent collaborations.

(c) Coordination of the First AstroPaleo Br Workshop in 2012 (http://www.astro-biobrasil.org/wapbr/WAPBr.pdf), organized in collaboration with the Institute of Geosciences. The event highlighted the application of new analytical techniques to common research problems and attracted approximately 70 participants.

(d) Hosting of the Second AstroPaleo Br Workshop in 2013 (www.astrobiobrasil. org/IIwapbr/), again in partnership with the Institute of Geosciences. This edition focused on both current and ancient extreme environments and included two international speakers, with around 80 participants in attendance.

(e) The creation of the Rede Brasileira de Astrobiologia (RBA, www.astrobiologia. net.br) in 2013 aimed to integrate the national scientific community and facilitate the exchange of research among NAP-Astrobio members, the Federal

University of São Carlos (UFSCar), the National Synchrotron Light Laboratory (LNLS), and other institutions across Brazil.

(f) The Second Brazilian Workshop on Astrobiology (http://www.astro.iag.usp. br/~2ndbwa) was held in 2013 in the city of Guarujá, bringing together approximately 70 researchers from across the country and featuring three international keynote speakers. This event was of considerable significance for the development of astrobiology in Brazil, fostering academic exchange and dialogue on global initiatives, including the launch of dedicated academic programs and the Brazilian Astrobiology Network (RBA).

(g) NAP-Astrobio's admission as an international affiliate member of both the NASA Astrobiology Institute (NAI) and the European Astrobiology Network Association (EANA) significantly elevated the international prestige of Brazilian astrobiology. This affiliation enabled researcher and student exchanges, including participation in global events such as the Astrobiology Graduate Conference (AbGradCon) in 2012 and the Santander Summer School in 2012 and 2013.

(h) NAP-Astrobio's academic and scientific contributions include the publication of the first scientific astrobiology book authored in Brazil, Astrobiologia—Uma Ciência Emergente (Galante et al., 2016), made freely available in PDF format (link). Additional outputs include a reference atlas on microbialites (Fairchild et al., 2015), a book chapter (Iten, 2016), over 76 peer-reviewed scientific articles, six scientific events, four master's theses, and three doctoral dissertations.

Currently, there is no undergraduate or graduate program specifically dedicated to astrobiology in Brazil; however, several universities offer courses and subjects that address related topics. The first course explicitly focused on the field was titled Bioastronomy, launched in 2001 at the Universidade do Vale dos Sinos (Unisinos) in southern Brazil (IHU, 2012). The following year, an Exobiology course was established for undergraduate biology students at the Universidade Federal do Rio Grande do Sul (UFRGS), coordinated by Jorge A. Quillfeldt in collaboration with faculty from diverse disciplines such as biology, physics, and geosciences. This course has covered foundational concepts in astrobiology and space exploration (Quillfeldt, 2010).

"Origens e Evolução do Universo, da Terra e da Vida" was the title of the course created by Dimas Zaia at the Universidade Estadual de Londrina (UEL) in 2003. It has since evolved into a formal specialization program comprising 360 instructional hours and is now officially titled Astrobiology.

In 2003, the Instituto de Astronomia, Geofísica e Ciências Atmosféricas (IAG/ USP) launched an undergraduate course titled Life in the Cosmic Context, designed for students in both the natural sciences and the humanities. Initially coordinated by Augusto Damineli and later by Amâncio Friaça and Eduardo Janot Pacheco, the course integrated content from astrophysics, biology, geology, meteorology, chemistry, and other scientific disciplines, embodying a transdisciplinary approach. By 2009, it had attracted students from 23 different academic specializations. In 2010, a panel discussion on astrobiology and biodiversity was held, during which participants explored the relationship between astrobiological research and the role of biodiversity in maintaining stable biospheres, as well as strategies for addressing anthropogenic pressures on the Earth's system and potential alien biospheres.

Several astrobiology-related courses were subsequently established in Brazil. These include offerings at the *Observatório do Valongo* of the *Universidade Federal do Rio de Janeiro*, and at the *Universidade do Vale do Paraíba* (UNIVALE). The *Núcleo de Astrofísica Teórica* (NAT) of the *Universidade Cruzeiro do Sul* (UNICSUL) implemented a postgraduate specialization course in Astronomy Education in 2010, which featured a module on Astrobiology and Exoplanets (Leme, 2013). Initially coordinated by Diego F. Gonçalves (IAG/USP) and later by Paulo R. Leme, this course has trained approximately 30 educators and science communicators.

A Professional Master's Degree in Astronomy Education (MPEA), which includes a dedicated astrobiology module, was launched in 2013 at the *Instituto de Astronomia, Geofísica e Ciências Atmosféricas* (USP). Currently coordinated by Amâncio Friaça, the program has conferred degrees upon nearly 30 professionals already active in astronomy education and public outreach.

An Advanced School of Astrobiology (SPASA 2011) was organized by the *Universidade de São Paulo* in 2011, bringing together more than 120 undergraduate and graduate students, as well as early-career researchers. The event fostered a multidisciplinary environment, encouraging collaboration and knowledge exchange across diverse scientific fields within the astrobiological domain.

Finally, it is important to highlight the growing interest among undergraduate students from various disciplines in the study of the origin and evolution of life. Requests for lectures at events such as the "Biology Week" are frequently made to members of NAP and their colleagues, and these are generally well received. The next generation of astrobiologists—who may identify themselves as such rather than as biologists, chemists, or physicists—is currently being formed, and will largely be a product of the activities carried out to date.

The success of the first workshop inspired the organization of the second Brazilian Workshop on Astrobiology, held in the city of Guarujá from September 23–27, 2013. The event addressed key topics such as habitability, exoplanets, interstellar chemistry, early Earth and the origins of life, limits of life and extremophiles, as well as science communication and education in astrobiology. The Scientific Organizing Committee was composed of Eduardo Janot Pacheco (IAG-USP), Amâncio Friaça (IAG-USP), Enio Frota da Silveira (PUC-Rio), Jorge E. Horvath (IAG-USP), Cláudia Lage (IBCCF-UFRJ), Vivian Helena Pelizzari (IO-USP), Gustavo Porto de Mello (OV-UFRJ), Jorge A. Quillfeldt (UFRGS), Marcelo da Rosa Alexandre (UFS), Carlos Alexandre Wuensche (INPE), and Dimas Zaia (UEL). Invited international speakers included Natalie Batalha (Kepler Mission, USA), Jean Duprat (Université d'Orsay, France), Hideo Hashizume (NIMS, Japan), Robert E. Johnson (University of Virginia, USA), and Eva Mateo Marti (CAB, Spain).

5.8   National Synchrotron Light Laboratory

The Laboratório Nacional de Luz Síncrotron—LNLS (www.lnls.cnpem.br), operates the only synchrotron light source in Latin America, offering unique infrastructure for advanced microscopic analysis of matter through infrared, ultraviolet, and X-ray radiation. As an open-access facility, the LNLS hosts approximately 1200 researchers annually from Brazil and abroad, supporting over 400 experimental projects that result in around 200 scientific publications each year.

The facility provides access to a broad spectral range—from vacuum ultraviolet (VUV) to soft X-rays (10–8000 eV)—which enables the simulation of the high-energy radiation environment that does not reach Earth's surface but plays a significant role in interstellar and planetary contexts. Beyond astrochemistry, LNLS beamlines are used to assess the survivability of microorganisms under space-like conditions, including simulated extraterrestrial surfaces and radiation exposure scenarios (Paulino-Lima et al., 2010, 2011; Abrevaya et al., 2011). These experiments can be further integrated with simulation chambers such as those at AstroLab, enhancing the capacity to investigate the limits of life and the stability of potential biosignatures under planetary analog conditions.

5.9   Laboratory of Radiation in Biology—LaRBio

This important laboratory was established and has been coordinated by Claudia Lage (Federal University of Rio de Janeiro—UFRJ) since 2011. The foundation of all its projects is centered on DNA as both a marker and the metabolic locus of the studied responses, with genetic information constituting the basis of cellular organization throughout the terrestrial biosphere. This makes it the primary focus of the group's research, explored from various perspectives.

The projects associated with the lab's main research line include:

(a) Survival strategies of extremophilic microorganisms under various biological extremes and simulated extraterrestrial environments.
(b) Prospection of enzymes and pigments of biotechnological interest, induced by stressors affecting the metabolism of extremophilic microorganisms.
(c) Characterization of spectral signatures in the infrared range of complex biomolecules, to enable their tracking in databases and in future observations of exoatmospheres in habitable environments.
(d) Monitoring of atmospheric biomarkers during stratospheric balloon flights.

The scientific and academic output of LaRBio includes approximately ten articles published in peer-reviewed journals, six undergraduate research projects, three master's theses, two doctoral dissertations, and one postdoctoral research project.

Among the most notable national collaborative projects are the following:

(a) Instituto Nacional de Estudos do Espaço (INEspaço), which supported the creation of AstroLab—home to the first simulation chamber for extraterrestrial environments in the Southern Hemisphere, known as AstroCam—in collaboration with the Institute of Astronomy, Geophysics, and Atmospheric Sciences (IAG/USP).
(b) Laboratório Nacional de Luz Síncrotron (LNLS), which provided instrumental support for high-energy beam irradiation experiments.
(c) Instituto Nacional de Metrologia (INMETRO), which has contributed to the modeling of previously uncharacterized proteins in Deinococcus radiodurans.
(d) Laboratório de Modelagem e Dinâmica Molecular (LMDM-UFRJ), which provides computational support for in silico characterization of unknown proteins in D. radiodurans.

Significant international collaborations include:

(a) The Department of Physics and Astronomy, Open University UK, under a Joint Project funded by the Royal Society.
(b) The Department of Genetics and Microbiology, *Pontificia Universidad Católica de Chile.*
(c) The Laboratoire Universitaire d'Astrophysique de Nice (LUAN), Université de Nice Sophia-Antipolis, under the project "Étude multi-technique des environnements stellaires: physico-chimie, processus évolutif, molécules prébiotiques", with mission support funded by the PICS/CNRS Program, France.

5.10 Other Recent and Ongoing Projects

In addition to the ongoing projects developed at the research centers mentioned earlier, several other noteworthy initiatives are currently underway across Brazil. One particularly relevant example is the work conducted at the Laboratory of Prebiotic Chemistry (LQP-UEL), led by Dr. Dimas A. M. Zaia at the Universidade Estadual de Londrina.

This laboratory focuses on the synthesis of minerals under prebiotic chemical conditions, environments likely representative of the Earth before the emergence of life. One of the key approaches adopted involves the use of artificial seawater, a medium believed to closely approximate the chemical composition of primordial oceans some 4.0 billion years ago (Zaia, 2012a, 2012b; Carneiro et al., 2013, 2017).

A major research emphasis is placed on the interaction between biomolecules—such as amino acids and nucleobases—and mineral surfaces. These interactions are of great interest because they may have played a role in the stabilization, concentration, and polymerization of organic molecules in early Earth environments, thereby facilitating the transition from chemistry to biology.

Experiments are conducted with these biomolecules dissolved in synthetic seawater and subsequently analyzed using a variety of spectroscopic techniques, including infrared (IR) and nuclear magnetic resonance (NMR) spectroscopy

(Anizelli et al., 2015, 2016; Villafañe-Barajas et al., 2018). The results contribute valuable insights into the potential catalytic role of minerals in prebiotic chemistry and the possible pathways that led to the origin of life.

In 2011, the research group led by Jorge L. Meléndez-Moreno (IAG/USP) launched a program aimed at searching for planets around solar twins, employing the HARPS spectrograph (High Accuracy Radial velocity Planet Searcher) at the La Silla Observatory, operated by the European Southern Observatory (ESO). The group's first discovery was a Jupiter twin—a planet with nearly the same mass and orbital distance from its host star as Jupiter has from the Sun (Bedell et al., 2015).

In 2017, the team announced the discovery of two additional planets: a super-Neptune and a super-Earth. Remarkably, the host star exhibits chemical signatures that reflect the presence of planets, offering compelling evidence for the dynamic evolution of planetary systems (Meléndez et al., 2017).

Moreover, the Sun has been shown to be deficient in refractory elements compared to the average composition of solar twins—stars with stellar parameters very similar to those of the Sun. These chemical anomalies are strongly correlated with the condensation temperatures of dust-forming elements in the protoplanetary disk. This deficiency may be linked to the formation of rocky material in the early Solar System, and possibly to the formation of the terrestrial planets themselves (Meléndez, 2009).

An atlas spanning from the Precambrian to the Recent Period, focused on microbialites in Brazil, was published by Thomas R. Fairchild (Institute of Geosciences—IG/USP) (Fairchild et al., 2015). Fairchild, along with Mirian L.F. Pacheco (Federal University of São Carlos—Sorocaba campus) and collaborators, has also conducted extensive research on the Brazilian Ediacaran fauna and its potential relevance to the broader evolution of the biosphere within an astrobiological framework (Fairchild, 2012).

5.11 Brazilian Astrobiology Society—SBAstrobio

At the XLI Annual Meeting of the Brazilian Astronomical Society, held in São Paulo on September 6, 2017, the Brazilian Astrobiology Society (SBAstrobio) was officially founded. A provisional board of directors was elected for a one-year term, with the initial task of legally establishing the new society. The board was composed as follows: President—Eduardo Janot Pacheco (USP); Vice President—Dimas Zaia (UELondrina); General Secretary—Enio Silveira (PUC-Rio); Education Secretary—Gustavo Porto de Mello (UFRJ); Outreach Secretary—Douglas Galante (LNLS); Secretary—Cristine Carneiro (Federal University of Western Bahia); Treasurer—Amaury Augusto de Almeida (USP).

The current number of registered members is close to 60, although it does not yet fully represent the broader Brazilian astrobiology community, which is highly diverse and extensive.

5.12 Conclusions

We do not claim to have covered all astrobiology-related activities in Brazil, as this would be an almost impossible task, and certainly some relevant researchers may have been left out of this document—for which we apologize.

There are, however, solid reasons to maintain a well-founded optimism regarding the development of astrobiology in Brazil. The scientific community is active, diverse, and engaged in research of significant relevance across multiple fields. Nevertheless, there remains a pressing need to deepen and expand the shared perspectives within this inherently transdisciplinary activity. Planned initiatives have been severely impacted by the well-known political and economic crisis, which has significantly restricted research funding. As a result, research groups have often had to resort to drastic measures in order to sustain their work.

Yet, the conviction endures that astrobiology will continue to grow and transform, asserting itself as an autonomous field that is at once symbiotic and inseparable from the many disciplines it encompasses. This is the message being passed on to the next generations—already nurtured in an intellectually stimulating and innovative environment.

Declarations

Competing Interests The authors declare no competing interests.

This chapter is an extended and updated version of a previously published article: Rodrigues et al. (2012). Astrobiology in Brazil: early history and perspectives. International Journal of Astrobiology, 11(4), 189–202.

References

Abrevaya, X. C., Paulino-Lima, I. G., Galante, D., Rodrigues, F., Mauas, P. J., Cortón, E., & Lage, C. D. A. S. (2011). Comparative survival analysis of Deinococcus radiodurans and the Haloarchaea Natrialba magadii and Haloferax volcanii exposed to vacuum ultraviolet irradiation. *Astrobiology, 11*(10), 1034–1040. https://doi.org/10.1089/ast.2011.0607

Albarran, G., Ramos, S., Negron Mendoza, A., & Collins, C. H. (1993). A plausible route for the synthesis of bio-organic compounds in the primitive earth from mineral salts. An overview. In *International conference on chemical evolution and the origin of life: Self-organization of the macromolecules of life, Trieste.*

Alves, D., & Fontanari, J. F. (1997). Error threshold in the evolution of diploid organisms. *Journal of Physics A: Mathematical and General, 30*(8), 2601–2607. https://doi.org/10.1088/0305-4470/30/8/009

Alves, D., Campos, P. R. A., Silva, A. T. C., & Fontanari, J. F. (2000). Group selection models in prebiotic evolution. *Physical Review E, 63*(1), 011911. https://doi.org/10.1103/PhysRevE.63.011911

Andrade, D. P. P., Boechat-Roberty, H. M., Pilling, S., da Silveira, E. F., & Rocco, M. L. M. (2009). Positive and negative ionic desorption from condensed formic acid photoexcited around the O 1s-edge: Relevance to cometary and planetary surfaces. *Surface Science, 603*(22), 3301–3306. https://doi.org/10.1016/j.susc.2009.09.020

Anizelli, P. R., Baú, J. P. T., Gomes, F. P., da Costa, A. C. S., Carneiro, C. E., Zaia, C. T. B., & Zaia, D. A. (2015). A prebiotic chemistry experiment on the adsorption of nucleic acids bases onto a natural zeolite. *Origins of Life and Evolution of Biospheres, 45*, 289–306. https://doi. org/10.1007/s11084-015-9401-1

Anizelli, P. R., Baú, J. P. T., Valezi, D. F., Canton, L. C., Carneiro, C. E. A., Di Mauro, E., da Costa, A. C. S., Galante, D., Braga, A. H., Rodrigues, F., Coronas, J., Casado-Coterillo, C., Zaia, C. T. B. V., & Zaia, D. A. M. (2016). Adenine interaction with and adsorption on Fe-ZSM-5 zeolites: A prebiotic chemistry study using different techniques. *Microporous and Mesoporous Materials, 226*, 493–504. https://doi.org/10.1016/j.micromeso.2016.02.004

Bedell, M., Meléndez, J., Bean, J. L., Ramírez, I., Asplund, M., Alves-Brito, A., Casagrande, L., Dreizler, S., Monroe, T., Spina, L., & Tucci Maia, M. (2015). The solar twin planet search. II. A Jupiter twin around a solar twin. *Astronomy & Astrophysics, 581*, id.A34. https://doi. org/10.1051/0004-6361/201525748

Boechat-Roberty, H. M., Pilling, S., & Santos, A. C. F. (2005). Destruction of formic acid by soft X-rays in star-forming regions. *Astronomy & Astrophysics, 438*(3), 915–922. https://doi. org/10.1051/0004-6361:20042588

Boechat-Roberty, H. M., Neves, R., Pilling, S., Lago, A. F., & De Souza, G. G. B. (2009). Dissociation of the benzene molecule by ultraviolet and soft X-rays in circumstellar environment. *Monthly Notices of the Royal Astronomical Society, 394*(2), 810–817. https://doi. org/10.1111/j.1365-2966.2008.14368.x

Carneiro, C. E., de Santana, H., Casado, C., Coronas, J., & Zaia, D. A. (2011). Adsorption of amino acids (Ala, Cys, His, Met) on zeolites: Fourier transform infrared and Raman spectroscopy investigations. *Astrobiology, 11*(5), 409–418. https://doi.org/10.1089/ast.2010.0521

Carneiro, C. E. A., Ivashita, F. F., de Souza Junior, I. G., de Souza, C. M. D., Paesano, A., Jr., da Costa, A. C. S., di Mauro, E., de Santana, H., Zaia, C. T. B. V., & Zaia, D. A. M. (2013). Synthesis of goethite in solutions of artificial seawater and amino acids: A prebiotic chemistry study. *International Journal of Astrobiology, 12*(2), 149–160. https://doi.org/10.1017/ S1473550413000013

Carneiro, C. E., Stabile, A. C., Gomes, F. P., da Costa, A. C., Zaia, C. T., & Zaia, D. A. (2017). Interaction, at ambient temperature and 80 C, between minerals and artificial seawaters resembling the present ocean composition and that of 4.0 billion years ago. *Origins of Life and Evolution of Biospheres, 47*, 323–343. https://doi.org/10.1007/s11084-016-9524-z

Cavalcanti, A. R. O., & Ferreira, R. (2001). On the relative contente of G,C bases in codons of amino acids corresponding to Class I and II aminoacyl-tRNA synthetases. *Origins of Life and Evolution of Biospheres, 31*(3), 257–269. https://doi.org/10.1023/A:1010639521100

Cavalcanti, A. R., Soares Leite, E., Neto, B. B., & Ferreira, R. (2004). On the classes of aminoacyl-tRNA synthetases, amino acids and the genetic code. *Origins of Life and Evolution of the Biosphere, 34*, 407–420. https://doi.org/10.1023/B:ORIG.0000029881.14519.42

Collins, C., Albarrán, G., & Collins, K. (2000). Irradiation promoted production of organic precursor species in inorganic solids on the prebiotic earth. In M. Akaboshi, N. Fujii, & R. Navarro-González (Eds.), *The role of radiation in the origin and evolution of Life* (pp. 143–153). Kyoto University Press.

de Oliveira, V. E., Castro, H. V., Edwards, H. G., & de Oliveira, L. F. C. (2010). Carotenes and carotenoids in natural biological samples: A Raman spectroscopic analysis. *Journal of Raman Spectroscopy, 41*(6), 642–650. https://doi.org/10.1002/jrs.2493

Duarte, R. T. D. (2010). *Micro-organismos em ambientes criogênicos: Gelo glacial, solos expostos por recuo de geleiras e permafrost polares*. PhD, Universidade de São paulo.

Duffard, R., Lazzaro, D., Licandro, J., De Sanctis, M. C., Capria, M. T., & Carvano, J. M. (2004). Mineralogical characterization of some basaltic asteroids in the neighborhood of (4) Vesta: First results. *Icarus, 171*(1), 120–132. https://doi.org/10.1016/j.icarus.2004.05.004

Edwards, H. G. M., Ellis-Evans, J. C., Newton, E. M., Little, S. J., De Oliveira, L. F. C., Hodgson, D., & Doran, P. T. (2002). Fourier-transform Raman spectroscopic studies of chronological change in stromatolitic cores from Antarctic lake sediments. *International Journal of Astrobiology, 1*(4), 325–331. https://doi.org/10.1017/S1473550403001253

Edwards, H. G., De Oliveira, L. F., Cockell, C. S., Ellis-Evans, J. C., & Wynn-Williams, D. D. (2004a). Raman spectroscopy of senescing snow algae: Pigmentation changes in an Antarctic cold desert extremophile. *International Journal of Astrobiology, 3*(2), 125–129. https://doi.org/10.1017/S1473550404002034

Edwards, H. G. M., Wynn-Williams, D. D., Little, S. J., de Oliveira, L. F. C., Cockell, C. S., & Ellis-Evans, J. C. (2004b). Stratified response to environmental stress in a polar lichen characterized with FT-Raman microscopic analysis. *Spectrochimica Acta Part A: Molecular and Biomolecular Spectroscopy, 60*(8–9), 2029–2033. https://doi.org/10.1016/j.saa.2003.10.021

Fairchild, T. R. (2012). Evolution of Precambrian life in the Brazilian geological record. *International Journal of Astrobiology, 11*, 309–323.

Fairchild, T. R., Rohn, R., & Dias Brito, D. (2015). *Microbialitos do Brasil do Pré-cambriano ao Recente: Um Atlas. Ed.* Rio Claro, São Paulo: UNESP-IGCE-UNESPetro. 392p.

Ferraz-Mello, S. (1964). Sur le problème de la pression de radiation solaire dans la Théorie des Satellites Artificiels. *Comptes Rendus de l'Académie des Sciences Paris, 258*, 463–466.

Ferraz-Mello, S. (1966). Recherches sur le mouvement des satellites galilieens de Jupiter. *Bulletin Astronomique, 1*(4), 287–330.

Ferraz-Mello, S. (1988). A semi-numerical expansion of the averaged disturbing function for some very-high-eccentricity orbits. *Celestial Mechanics and Dynamical Astronomy, 45*(1–3), 65–68. https://doi.org/10.1007/978-94-009-0985-4_10

Ferraz-Mello, S., et al. (1993). On symmetrical planetary corotations. *Celestial Mechanics and Dynamical Astronomy, 55*(1), 25–45. https://doi.org/10.1007/BF00694393

Ferreira, R., & Cavalcanti, A. R. D. (1997). Vestiges of early molecular processes leading to the genetic code. *Origins of Life and Evolution of the Biosphere, 27*(4), 397–403. https://doi.org/1 0.1023/A:1006531904713

Ferreira, R., & Coutinho, K. R. (1993). Simulation studies of self-replicating oligoribotides, with a proposal for the transition to a peptide-assisted stage. *Journal of Theoretical Biology, 164*(3), 291–305. https://doi.org/10.1006/jtbi.1993.1155

Ferreira, R., & Tsallis, C. (1985). On the role of complementarity in biogenesis—A critical phenomenon approach. *Journal of Theoretical Biology, 117*(2), 303–317. https://doi.org/10.1016/ S0022-5193(85)80224-1

Ferreira-Rodrigues, A. M., et al. (2011). Photostability of amino acids to Lyman alpha radiation: Glycine. *International Journal of Mass Spectrometry, 306*(1), 77–81. https://doi.org/10.1016/j. ijms.2011.06.022

Fontanari, J. F., Santos, M., & Szathmáry, E. (2006). Coexistence and error propagation in prebiotic vesicle models: A group selection approach. *Journal of Theoretical Biology, 239*(2), 247–256. https://doi.org/10.1016/j.jtbi.2005.08.039

Friaça, A. C. S., & Terlevich, R. J. (1998). Formation and evolution of elliptical galaxies and QSO activity. *Monthly Notices of the Royal Astronomical Society, 298*(2), 399–415. https://doi. org/10.1046/j.1365-8711.1998.01626.x

Galante, D. (2009). *Efeitos astrofísicos e astrobiológicos de Gamma-Ray Bursts.* PhD, Universidade de São Paulo.

Galante, D., Silva, E. P. D., Rodrigues, F., Horvath, J. E., & Avellar, M. G. B. (2016). *Astrobiologia: Uma ciência emergente.* Núcleo de Pesquisa em Astrobiologia.

IHU. (2012). *Há vida fora da terra? As contribuições da exobiologia.* Retrieved April 18, 2025, from https://www.ihuonline.unisinos.br//media/pdf/IHUOnlineEdicao172.pdf

Iten, H. V. (2016). Origin and early diversification of phylum Cnidaria: Key macrofossils from the Ediacaran system of North and South America. In S. Goffredo & Z. Dubinsky (Eds.), *The Cnidaria, past, present and future—The world of Medusa and her sisters.* Springer.

Kuhn, E., Bellicanta, G. S., & Pellizari, V. H. (2009). New alk genes detected in Antarctic marine sediments. *Environmental Microbiology, 11*(3), 669–673. https://doi. org/10.1111/j.1462-2920.2008.01843.x

Lago, A. F., Coutinho, L. H., Marinho, R. R. T., De Brito, A. N., & De Souza, G. G. B. (2004). Ionic dissociation of glycine, alanine, valine and proline as induced by VUV (21.21 eV) photons. *Chemical Physics, 307*(1), 9–14. https://doi.org/10.1016/j.chemphys.2004.06.052

Léger, A., Rouan, D., Schneider, J., Barge, P., Fridlund, M., Samuel, B., Ollivier, M., Guenther, E., Deleuil, M., Deeg, H. J., Auvergne, M., Alonso, R., Aigrain, S., Alapini, A., Almenara, J. M., Baglin, A., Barbieri, M., Bruntt, H., Bordé, P., Bouchy, F., Cabrera, J., Catala, C., Carone, L., Carpano, S., Csizmadia, S., Dvorak, R., Erikson, A., Ferraz-Mello, S., Foing, B., & Fressin, F. (2009). Transiting exoplanets from the CoRoT space mission VIII. CoRoT-7b: The first super-Earth with measured radius. *Astronomy & Astrophysics, 506*(1), 287–302. https://doi.org/10.1051/0004-6361/200911933

Leme, P. R. (2013). *Graduate Certificated Course on Astrobiology and Exoplanets at Universidade Cruzeiro do Sul*. Presented as poster in 2nd Brazilian Workshop on Astrobiology, Guarujá, Brazil.

Lins, R., Soares, T. A., & Ferreira, R. (1996). Plural origins of the molecular homochirality in our biota. *Zeitschrift für Naturforschung C, 51*(1–2), 70–74.

Maciel, W. J., Costa, R. D. D., & Uchida, M. M. M. (2003). An estimate of the time variation of the O/H radial gradient from planetary nebulae. *Astronomy & Astrophysics, 397*(2), 667–674. https://doi.org/10.1051/0004-6361:20021530

Meléndez, J. (2009). The peculiar solar composition and its possible relation to planet formation. *The Astrophysical Journal Letters, 704*(1), L66–L70. https://doi.org/10.1088/0004-637X/704/1/L66

Meléndez, J., Bedell, M., Bean, J. L., Ramírez, I., Asplund, M., Dreizler, S., Yan, H.-L., Shi, J.-R., Lind, K., Ferraz-Mello, S., Yana Galarza, J., dos Santos, L., Spina, L., Tucci Maia, M., Alves-Brito, A., Monroe, T., & Casagrande, L. (2017). The Solar Twin Planet Search. V. Close-in, low-mass planet candidates and evidence of planet accretion in the solar twin HIP 68468. *Astronomy & Astrophysics, 597*, id.A34. https://doi.org/10.1051/0004-6361/201527775

Michtchenko, T., & de Mello, G. F. P. (2009). Dynamical Stability of Habitable Planets in Astrobiologically Interesting Binary Stars. *Bioastronomy 2007: Molecules, Microbes, and Extraterrestrial Life, 420*, 353–356.

Michtchenko, T. A., et al. (2002). Origin of the basaltic asteroid 1459 Magnya: A dynamical and mineralogical study of the outer main belt. *Icarus, 158*(2), 343–359. https://doi.org/10.1006/icar.2002.6871

Miller, S. L. (1953). A production of amino acids under possible primitive earth conditions. *Science, 117*(3046), 528–529. https://doi.org/10.1126/science.117.3046.528

Nesvorny, D., & Ferraz-Mello, S. (1997). On the asteroidal population of the first-order Jovian resonances. *Icarus, 130*(2), 247–258. https://doi.org/10.1006/icar.1997.5807

Neves, R., et al. (2007). Photodissociation of deutered benzene by soft X-ray: Astrophysical implications. *Journal of Electron Spectroscopy and Related Phenomena, 156*, LXXIII.

Paulino-Lima, I. G. (2010). *Investigação das condições de sobrevivência de microrganismos extremófilos em ambientes extraterrestres simulados*. PhD, Universidade Federal do Rio de janeiro.

Paulino-Lima, I. G., Pilling, S., Janot-Pacheco, E., de Brito, A. N., Gonçalves Barbosa, J. A. R., Leitão, A. C., & de Alencar Santos Lage, C. (2010). Laboratory simulation of interplanetary ultraviolet radiation (broad spectrum) and its effects on Deinococcus radiodurans. Planet. *Planetary and Space Science, 58*(10), 1180–1187. https://doi.org/10.1016/j.pss.2010.04.010

Paulino-Lima, I. G., Janot-Pacheco, E., Galante, D., Cockell, C., Olsson-Francis, K., Brucato, J. R., Baratta, G. A., Strazzulla, G., Merrigan, T., McCullough, R., Mason, N., & Lage, C. (2011). Survival of deinococcus radiodurans against laboratory-simulated solar wind charged particles. *Astrobiology, 11*(9), 875–882. https://doi.org/10.1089/ast.2011.0649

Pereira, F. A. (1958). *Introdução à Astrobiologia*. José Olympio Editora.

Picazzio, E., de Almeida, A. A., Andrievskii, S. M., Churyumov, K. I., & Luk'yanyk, I. V. (2007). A high spectral resolution atlas and catalogue of emission lines of the comet C/2000

WM1 (LINEAR). *Advances in Space Research, 39*(3), 462–467. https://doi.org/10.1016/j.asr.2003.06.048

Pilling, S., Santos, A. C. F., & Boechat-Roberty, H. M. (2006a). Photodissociation of organic molecules in star-forming regions-II. Acetic acid. *Astronomy & Astrophysics, 449*(3), 1289–1296. https://doi.org/10.1051/0004-6361:20053927

Pilling, S., Santos, A. C. F., Wolff, W., Sant'Anna, M. M., Barros, A. L. F., De Souza, G. G. B., et al. (2006b). Ionization and dissociation of cometary gaseous organic molecules by solar wind particles-I. Formic acid. *Monthly Notices of the Royal Astronomical Society, 372*(3), 1379–1388. https://doi.org/10.1111/j.1365-2966.2006.10949.x

Pilling, S., Neves, R., Santos, A. C. F., & Boechat-Roberty, H. M. (2007). Photodissociation of organic molecules in star-forming regions-III. Methanol. *Astronomy & Astrophysics, 464*(1), 393–398. https://doi.org/10.1051/0004-6361:20066275

Pilling, S., Andrade, D. P., Neto, A. C., Rittner, R., & Naves de Brito, A. (2009). DNA nucleobase synthesis at Titan atmosphere analog by soft X-rays. *The Journal of Physical Chemistry A, 113*(42), 11161–11166.

Quillfeldt, J. (2010). *Disciplina BIO10–012: Exobiologia*. Retrieved April 13, 2025, from http://exobiologia.ufrgs.br/

Rocha-Pinto, H. J., Flynn, C., Scalo, J., Hänninen, J., Maciel, W. J., & Hensler, G. (2004). Chemical enrichment and star formation in the Milky Way disk-III. Chemodynamical constraints. *Astronomy & Astrophysics, 423*(2), 517–535. https://doi.org/10.1051/0004-6361:20035617

Rodrigues, D. F., Jesus, E. D. C., Ayala-del-Río, H. L., Pellizari, V. H., Gilichinsky, D., Sepulveda-Torres, L., & Tiedje, J. M. (2009). Biogeography of two cold-adapted genera: Psychrobacter and Exiguobacterium. *The ISME Journal, 3*(6), 658–665. https://doi.org/10.1038/ismej.2009.25

Rodrigues, F., Galante, D., Paulino-Lima, I. G., Duarte, R. T., Friaça, A. C., Lage, C., Janot-Pacheco, E., Teixeira, R., & Horvath, J. E. (2012). Astrobiology in Brazil: Early history and perspectives. *International Journal of Astrobiology, 11*(4), 189–202. https://doi.org/10.1017/S1473550412000250

Silvestre, D. G., & Fontanari, J. F. (2005). Template coexistence in prebiotic vesicle models. *The European Physical Journal B-Condensed Matter and Complex Systems, 47*, 423–429. https://doi.org/10.1140/epjb/e2005-00346-5

Soares, T. A., et al. (1997). Plural origins of molecular homochirality in our biota 0.2. The relative stabilities of homochiral and mixed oligoribotides and peptides. *Zeitschrift für Naturforschung C: Journal of Biosciences, 52*(1–2), 89–96.

Teixeira, L. C., Peixoto, R. S., Cury, J. C., Sul, W. J., Pellizari, V. H., Tiedje, J., & Rosado, A. S. (2010). Bacterial diversity in rhizosphere soil from Antarctic vascular plants of Admiralty Bay, maritime Antarctica. *The ISME Journal, 4*(8), 989–1001. https://doi.org/10.1038/ismej.2010.35

Tessis, A. C., De Amorim, H. S., Farina, M., De Souza-Barros, F., & Vieyra, A. (1995). Adsorption of 5′-AMP and catalytic synthesis of 5′-ADP onto phosphate surfaces: Correlation to solid matrix structures. *Origins of Life and Evolution of the Biosphere, 25*, 351–373. https://doi.org/10.1007/BF01581775

Tsallis, C. (2008). My friend Ricardo Ferreira, an impressive natural philosopher. *Journal of the Brazilian Chemical Society, 19*, 203–205. https://doi.org/10.1590/S0103-50532008000200002

Tsallis, C., & Ferreira, R. (1983). On the origin of self-replicating information-containing polymers from oligomeric mixtures. *Physics Letters A, 99*(9), 461–463. https://doi.org/10.1016/0375-9601(83)90958-1

Vieyra, A., Gueiros-Filho, F., Meyer-Fernandes, J. R., Costa-Sarmento, G., & De Souza-Barros, F. (1995). Reactions involving carbamyl phosphate in the presence of precipitated calcium phosphate with formation of pyrophosphate: A model for primitive energy-conservation pathways. *Origins of Life and Evolution of the Biosphere, 25*, 335–350. https://doi.org/10.1007/BF01581774

Villafañe-Barajas, S. A., Baú, J. P. T., Colín-García, M., Negrón-Mendoza, A., Heredia-Barbero, A., Pi-Puig, T., & Zaia, D. A. (2018). Salinity effects on the adsorption of nucleic acid compounds on Na-montmorillonite: A prebiotic chemistry experiment. *Origins of Life and Evolution of Biospheres, 48,* 181–200. https://doi.org/10.1007/s11084-018-9554-9

Zaia, D. A. (2003). From spontaneous generation to prebiotic chemistry. *Quimica Nova, 26,* 260–264.

Zaia, D. A. M. (2004). A review of adsorption of amino acids on minerals: Was it important for origin of life? *Amino Acids, 27*(1), 113–118. https://doi.org/10.1007/s00726-004-0106-4

Zaia, D. A. M. (2012a). *Website of the discipline Astrobiologia—Universidade estadual de Londrina.* Retrieved April 20, 2025, from http://www.uel.br/pos/astrobiologia/

Zaia, D. A. M. (2012b). Adsorption of amino acids and nucleic acid bases onto minerals: A few suggestions for prebiotic chemistry experiments. *International Journal of Astrobiology, 11*(4), 229–234. https://doi.org/10.1017/S1473550412000195

Zaia, D. A., Vieira, H. J., & Zaia, C. T. (2002). Adsorption of L-amino acids on sea sand. *Journal of the Brazilian Chemical Society, 13,* 679–681.

Zaia, D. A., Zaia, C. T. B., & De Santana, H. (2008). Which amino acids should be used in prebiotic chemistry studies? *Origins of Life and Evolution of Biospheres, 38,* 469–488. https://doi.org/10.1007/s11084-008-9150-5

Chapter 6
Astrobiology and Planetary Sciences in Chile

Giovanni Leone, Rómulo Oses, and Ricardo Cabrera

Abstract This chapter provides a comprehensive overview of the current landscape of astrobiology and planetary sciences in Chile, highlighting its scientific achievements, institutional developments, and future prospects. With the Atacama Desert recognized as one of the most valuable Mars analogue sites on Earth, Chile has emerged as a key player in international astrobiology research, contributing to major initiatives such as ARADS and ExoFiT. These projects have tested technologies for biosignature detection, contamination control, and rover operations under Mars-like conditions. The chapter also explores the growing field of planetary sciences in Chile, which benefits from an extensive network of observatories and a rising interest in Solar System studies, including CubeSat development, meteoritics, and lunar base simulations. While institutional representation in international collaborations remains limited, recent efforts such as the Latin American Artemis Project Workshop (LAAP24) and the strengthening of academic programs suggest a promising trajectory. Public outreach initiatives and the incorporation of planetary topics into university curricula have contributed to increasing student interest. Overall, Chile's unique natural laboratories, scientific expertise, and strategic initiatives position it to play an increasingly central role in advancing space science and contributing to the objectives of international frameworks such as the Artemis Accords.

G. Leone (✉)
Instituto de Investigación en Astronomía y Ciencias Planetarias, Universidad de Atacama, Copiapó, Chile
e-mail: giovanni.leone@uda.cl

R. Oses
Centro Regional de Investigación y Desarrollo Sustentable de Atacama (CRIDESAT), Universidad de Atacama, Copiapó, Chile

R. Cabrera
Departamento de Biología, Facultad de Ciencias, Universidad de Chile, Santiago, Chile

D. Tovar, M. A. Leal (eds.), *Astrobiology and Planetary Sciences in Latin America*, https://doi.org/10.1007/978-3-032-01450-4_6

6.1 Introduction

Chile, also thanks to its university accreditation system (Gerón-Piñón et al., 2021; Lamarra, 2003), has made important qualitative steps forward in terms of research and it is building a vibrant community covering almost every field of science. Among the countries of Latin America, Chile invested a significant amount of money to improve its academic and scientific environment (Véliz & Celis, 2021). As a result, there is a wealth of universities all over the country with a significant concentration in its capital Santiago.

Although Chile has made significant progress in almost every field of science, we focus in this chapter on the fields of Astrobiology and Planetary Sciences. For their intrinsic complexity and integrative approach (their studies also involve the more general fields of Astronomy, Biology, and Geology) both are considered as multi-inter-transdisciplinary sciences (Fishbaugh et al., 2007; Santos et al., 2016). Both fields are intimately interconnected in every aspect of life's interaction with the planetary environment. Rather than focusing primarily on their scientific aspects, in this chapter we will explore the overall situation of Astrobiology and Planetary Sciences research, education, and public outreach in Chile. This is particularly interesting in view of the involvement of several Latin American countries in the Artemis Accords (Cocca y Esquivel, 2024), an international agreement aimed at going back to the Moon and using it as a launching pad for Mars, and useful to understand how these countries may contribute to the international efforts for the return to the Moon.

6.2 Current Situation

6.2.1 Astrobiology

Astrobiology in Chile has developed significantly around the exceptional characteristics of the Atacama Desert, which serves as one of Earth's premier Mars analogue sites. The Atacama's extreme environmental conditions—high UV radiation levels and exceptional aridity—create a natural laboratory that resembles just a part of the Martian surface conditions (Aerts et al., 2020; Veneranda et al., 2021). Clearly, the overall environment of Mars is significantly different from that of the Earth, although geomorphologically speaking some places of the Atacama Desert are quite similar to those on Mars (Fig. 6.1).

While the available literature does not provide comprehensive details about the historical origins of astrobiology as a formal discipline in Chile, evidence shows that international scientific interest in the region's astrobiological significance has intensified over the past two decades. The development of astrobiology research in Chile appears to have evolved alongside the growing global interest in Mars exploration missions. The Atacama Desert is becoming increasingly recognised for its value in testing technologies and methodologies intended for deployment on Mars

Fig. 6.1 Geomorphological comparison between Mars and the Earth; (a) Sector Caseron, Atacama Desert (along the motorway between Caldera and Copiapò, Chile); (b) Gale Crater, Mars

(Glass et al., 2023). This trajectory has positioned Chile as a critical contributor to planetary science and astrobiology despite challenges in local institutional representation and their possible involvement in international research initiatives (Tavernier et al., 2023).

The Atacama Desert, particularly its hyperarid core, represents one of Earth's most significant Mars analogue environments. Its extreme conditions make it an exceptional location for studying the limits of life and testing technologies for future Mars missions (Aerts et al., 2020; Glass et al., 2023). The desert's value as an astrobiology research site stems from several key characteristics:

1. *Extreme Aridity*: Parts of the Atacama are among the driest places on Earth, receiving negligible precipitation and featuring soil conditions that parallel Mars's oxidising, low-organic-matter environment (Aerts et al., 2020).
2. *High UV Radiation*: The combination of high elevation, atmospheric clarity, and low latitude results in intense solar radiation similar to Martian surface conditions (Veneranda et al., 2021; Shen et al., 2022; Abrahamsson & Kanik, 2022).
3. *Distinctive Salt Flats (Salars):* These environments host extremophile communities that may provide insights into potential survival mechanisms for life in hostile extraterrestrial environments (Aerts et al., 2020).
4. *Geochemical Composition:* The mineral composition of Atacama soils contains analogues to Martian regolith, including various sulphates and chlorides that influence potential habitability (Aerts et al., 2020; Veneranda et al., 2021).

These characteristics collectively make the Atacama an invaluable natural laboratory for astrobiology research. It allows scientists to investigate fundamental questions about the limits of life and develop technologies for detecting potential biosignatures on Mars (Bonaccorsi et al., 2023; Glass et al., 2023; Parro et al., 2018) (Fig. 6.2).

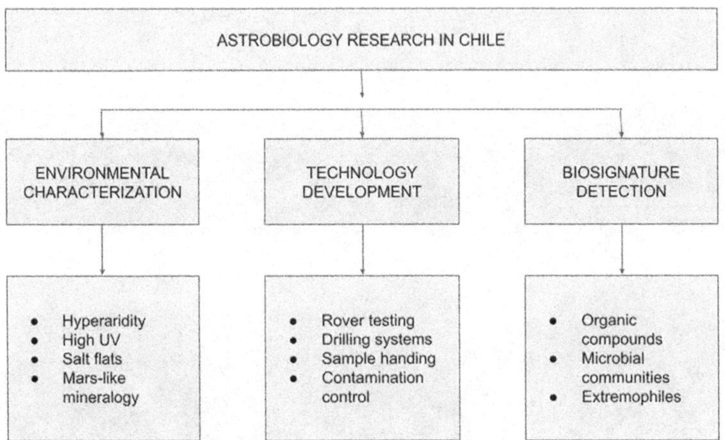

Fig. 6.2 Conceptual Framework of Astrobiology Research in the Atacama Desert

6.2.2 Major Research Initiatives and Technological Developments

1. *Atacama Rover Astrobiology Drilling Studies (ARADS)*

The ARADS project represents one of the most significant astrobiology research initiatives in Chile's Atacama Desert. This NASA-led project tested an autonomous rover-mounted robotic drill prototype designed for a potential Mars life detection mission (Glass et al., 2023; Bonaccorsi et al., 2023). ARADS focused on developing sample collection and handling protocols in Mars-like environments, emphasising contamination control—a critical factor for credible life detection missions (Bonaccorsi et al., 2023).

In 2019, ARADS field-tested this system during a simulated 6-Sol Mars mission, implementing sophisticated contamination control strategies essential for astrobiology investigations (Bonaccorsi et al., 2023). The project developed a comprehensive Contamination Control Strategy and Implementation (CCSI) for the Sample Handling and Transfer System (SHTS) hardware, which included:

(a) A five-step decontamination protocol combining hydrogen peroxide and sodium hypochlorite sterilants

(b) *In situ* real-time verification methods using adenosine triphosphate (ATP) detection

(c) Signs of Life Detector (SOLID) Fluorescence Immunoassay for bioburden characterisation

This protocol achieved a 4-log bioburden reduction, demonstrating the feasibility of maintaining aseptic conditions during drilling operations in Mars-like environments—a critical capability for future Mars sample collection missions (Bonaccorsi et al., 2023).

2. *ExoFiT Trials and Raman Spectroscopy Applications*

Another significant research initiative was the ExoFiT (Exomars-Preparation Functional Testing) trial, which involved the remote operation of a Rosalind Franklin rover emulator in the Atacama Desert (Veneranda et al., 2021; Tavernier et al., 2023). This project simulated the operational sequence of the ExoMars mission, culminating in the collection and analysis of drilled core samples (Veneranda et al., 2021; Tavernier et al., 2023; Abrahamsson & Kanik, 2022).

A key analytical component of the ExoFiT trial was the deployment of the Raman Laser Spectrometer (RLS) technology demonstrator (RAD1 system) for *in situ* sample characterisation (Veneranda et al., 2021; Tavernier et al., 2023; Abrahamsson & Kanik, 2022). The Raman systems successfully:

(a) Identified the mineralogical composition of drilled cores (k-feldspar, plagioclase, quartz, muscovite, and rutile)

(b) Detected minor mineral phases below the detection limit of X-ray diffractometry

(c) Identified organic functional groups ($-C \equiv N$, $-NH_2$, and $C-(NO_2)$) indicating the presence of nitrogen-fixing microorganisms

The detection of organic material in the subsurface of this Mars-analogue environment provides compelling evidence that Raman spectroscopy can effectively identify potential biosignatures in similar environments on Mars, supporting its inclusion in future missions (Veneranda et al., 2021; Tavernier et al., 2023; Abrahamsson & Kanik, 2022).

3. *Biosignature Detection in Extreme Environments*

A primary focus of astrobiology research in Chile has been developing and testing methodologies for detecting biosignatures in extreme environments. Researchers have conducted detailed analyses of the organic content in samples from various locations in the Atacama, including different wet and dry salt flats at intermediate to high elevations (Fig. 6.3).

These studies have employed sophisticated analytical techniques, including:

(a) Liquid Chromatography and Multidimensional Gas Chromatography-Mass Spectrometry for detecting and analysing amino acids

(b) Advanced extraction procedures to isolate potential biosignatures

(c) Illumina 16S amplicon sequencing for identifying microbial communities (Aerts et al., 2020)

While these analyses generally revealed low amino acid loads and minimal organic carbon and nitrogen quantities, they identified complex microbial communities dominated by extremophilic organisms, particularly halophilic microorganisms from the Archaeal family Halobacteriaceae. These findings demonstrate that life can persist even in environments dominated by halites and sulphates, suggesting parallel possibilities for potentially habitable niches on Mars (Aerts et al., 2020; Parro et al., 2011).

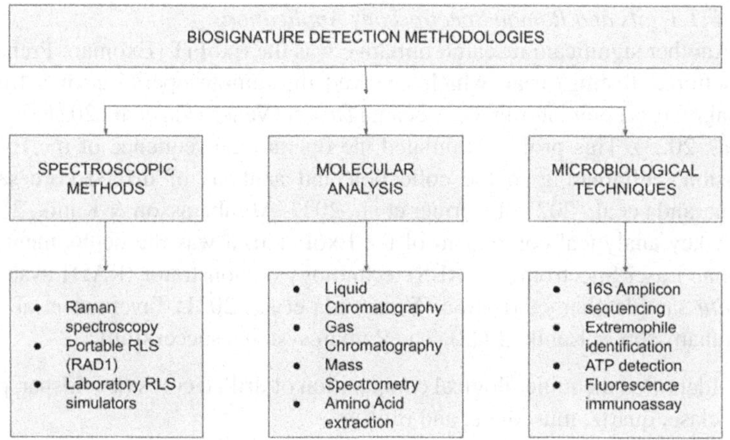

Fig. 6.3 Methodological Framework for Biosignature Detection in Chilean Astrobiology Research

6.2.3 Planetary Sciences

The field of Planetary Sciences in Chile is mostly related to the large number of telescopes that allow the observation of planetary systems in our galaxy. However, Planetary Sciences in Chile are not developed with telescopes only, we will show significant developments in Solar System studies and development of space technology. The specification of the Solar System subfield studies has been made because Planetary Sciences studies were already existing in Chile but mainly in the subfield of Exoplanets with the works of James Jenkins (Jenkins et al., 2017, 2019, 2020) and Patricio Rojo (Cortés-Zuleta et al., 2020; Jones et al., 2014; Hoyer et al., 2011), and in the subfield of Protoplanetary Disks with the works of Simon Casassus and Lucas Cieza (Pérez et al., 2019; Casassus et al., 2015; Cieza et al., 2015; Montesinos et al., 2015; Kraus et al., 2013) among others.

Regarding the Exoplanets, the classic methods are those of the transit survey and radial velocity but also direct imaging of exoplanets through sensitive adaptive optics, a technique particularly used in the search for exomoons (Lukić, 2017). The Universidad Diego Portales (UDP) has a rich program of research for the characterization of cometary dust, studies of protoplanetary disks, and a new program for the exploration of faint objects in the Outer Solar System (OSS) through the Vera Rubin Observatory (VRO). The Astronomy Centre of the Antofagasta University (UA) includes minor bodies of the Solar System and exoplanets among their lines of research.[1] The Astrophysics Institute of the Pontificia Católica Universidad de Chile (PUC)[2] mentions Viviana Guzman for protoplanetary disks (Guzmán et al., 2021) and Cristobal Petrovich for exoplanets (Petrovich et al., 2020). The Universidad

---

[1] https://www.astro.uantof.cl/research/lineas/

[2] https://astronomia.uc.cl/2jucext1/institutoastrofisica/en/investigacion-2/investigation

Andres Bello (UAB) studies the structure of the Exoplanets within the broader area of Stellar Astrophysics with an interesting work on a qualitative classification of extraterrestrial civilizations that involves Claudio Caceres and Dante Minniti (Ivanov et al., 2020).

Regarding the specific Solar System science as a subfield of the more general Planetary Sciences field, well, it is still in development in Chile although with the first significant results. The University of Chile (UCH) and the University of Santiago de Chile (USACH) have a joint collaboration on space physics and space instrumentation through the Space and Planetary Exploration Laboratory (SPEL) built in 2011 in the Electric Engineering Department of the UCH for the design, construction, and development of the first (nano)satellite of the University of Chile for Aerospace Investigation (SUCHAI).[3] The Universidad Católica del Norte (UCN) has vibrant research in meteoritics and geoheritage legal protection of Chilean impact craters thanks to the leading work of Millarca Valenzuela (Pinto et al., 2024; Valenzuela & Benado, 2018; Hutzler et al., 2016; Gattacceca et al., 2011). Even the Universidad de Atacama (UDA) experienced a growth in scientific production and the birth of both Astrobiology and Planetary Sciences fields among its academic ranks. In the years 2017–2020, a platoon of 27 researchers was hired by the UDA within the framework of a project to increase the university's scarce scientific production. This was a particularly important investment for a small regional university like UDA. Specific examples were the hiring of Romulo Oses, who previously worked at the Universidad Católica del Norte (UCN) (Molina-Montenegro et al., 2016a, 2016b; Torres-Díaz et al., 2016), for the field of Astrobiology, Giovanni Leone, who previously worked at Eidgenössische Technische Hochschule (ETH) Zürich (Leone, 2014, 2016, 2017; Leone et al., 2014; de Pater et al., 2014), for the specific studies of Solar System in the Planetary Sciences field, and Jeremy Tregloan-Reed, who worked as NASA Postdoctoral Fellow (Tregloan-Reed et al., 2015), for the field of Exoplanets studies. Other than sharing its name with the Atacama Desert, UDA is strategically located in its southern edge within an area of naturalistic interest.

We have already shown how the intimate relationship between Astrobiology and Mars studies in Chile has been developed in the Atacama Desert. The Atacama Desert is worldwide known as a natural laboratory for the fields of Astrobiology (Oses Pedraza et al., 2023; Krone et al., 2021) and Planetary Sciences (San Martin et al., 2025), which hosted experiments and tests with planetary rovers (Glass et al., 2023; Wei et al., 2015), and it will be a great opportunity of scientific production at international level aimed at the involvement of Chile in the Artemis Accords.[4] A significant contribution comes from the UDA and it is the characterization of potential landing sites where to deploy the first (human and/or robotic) settlement on the

---

[3] https://spel.ing.uchile.cl/

[4] https://www.nasa.gov/news-release/nasa-welcomes-chile-as-newest-artemis-accords-signatory/#:~:text=The%20Republic%20of%20Chile%20is%20the%2047th%20country,cooperation%20among%20nations%20participating%20in%20NASA%E2%80%99 s%20Artemis%20program

Moon, the Sverdrup-Henson crater has been proposed (Leone et al., 2023). This was still an individual effort rather than a systemic one, although in the framework of an international collaboration that also involves a NASA scientist but not yet Latin American institutions as a whole. Actually, local Chilean institutions are still quite underrepresented in Planetary Science research related to Atacama Desert (Tavernier et al., 2023). Therefore, as well as in the case for the organized efforts aimed at promoting Astrobiology in Brazil, the same effort has been done for Solar System studies in Latin America with a first workshop dedicated to the Artemis Project organized in Santiago last November 4, 2024, at the PUC[5] (Latin American Artemis Project 2024—LAAP24). This workshop was aimed at creating a multidisciplinary meeting point for Latin American public and private institutions, including private companies, in order to exchange scientific ideas and cooperate towards the fulfilment of the objectives stated in the Artemis Accords. On the other hand, the workshop "Astrobiology in Chile 2024" was held in San Esteban, Valparaíso from August 26 to 28, 2024. This workshop aimed to bring together the Chilean scientific community, promote interdisciplinary dialogue, and provide a space for collaboration and national-level coordination. Topics covered at this event included: (1) habitable planetary systems, (2) chemistry before planets, (3) from prebiotic chemistry to the evolution of life, and (4) the limits of life.

6.3 Perspectives of Research

Despite the general lack of funding, which affects more the public universities rather than the private ones, and the ever-increasing article processing costs (APC) for open access (OA) publication, which are dramatically draining important financial resources from research efforts in developing countries like Chile that do not benefit of publication fees waiver programs (Krauskopf, 2021), the scientific production at Chilean institutions is keeping pace with the growth of its last years thanks to the (ever-)decreasing number of journals that keep the vital free subscription option for publication. Some universities like UCH and PUC are more capable than others, also due to their higher numbers of researchers, in attracting public funds for research in Astronomy (although much less for Planetary Sciences) from the National Agency for Research and Development (Agencia Nacional de Investigación y Desarrollo—ANID in Spanish). Obviously, as will be shown in the next section, research in Astronomy and Astrophysics is predominant with respect to Planetary Sciences and thus this situation affects the amount of funding and the capability to publish significant research both in subscription and OA mode. Projects to sustain research in Planetary Sciences indeed do exist, such as the "Núcleo Milenio de Formación Planetaria,"[6] but once more they go towards the research in

---

[5] https://www.youtube.com/watch?v=W-YfdhiZ3Ow

[6] https://www.iniciativamilenio.cl/NPF/

Exoplanets rather than Solar System. Exceptions are the projects developed by Millarca Valenzuela at UCN regarding geoheritage and meteorite collection (Valenzuela & Benado, 2018), the first Chilean planetary simulant (CAD-1) found in the Atacama Desert by José San Martin Lobos and Giovanni Leone (San Martin et al., 2025), the development of CubeSat technologies at Universidad Técnica Federico Santa María (USM) by Rodrigo Cassinelli Palharini (Caqueo et al., 2023, 2024; Mutis et al., 2024), and another interesting project led by Grace Batalla and Lucas Cieza at UDP that studies Atacama chondrites olivine spectra comparing them to those in protoplanetary disks (Batalla Falcón, 2020; Batalla-Falcon et al., 2025).

Also, the LAAP24 has shown how a starting small base of actors that can play an important role in the build-up of a scientific and infrastructural spatial ecosystem already exists in Chile. The Chilean Air Force (Fuerza Aérea de Chile—FACH) is building its new headquarters in Santiago and is posed to establish the new Chilean Space Agency (Agencia Espacial Chilena—AEC) that will manage all the operations and space policy of Chile thanks to its own funding. Regarding the space policy and the related 1967 Outer Space Treaty, a group of attorneys and engineers specialized in space law constituted the Space Association of Chile (Asociación Chilena del Espacio—ACHIDE) led by Loreto Moraga[7] and can play a counselling role for the FACH-AEC as well as other institutions involved in space research and development (R&D). Indeed, ACHIDE already proposed the current international management of Antarctica as a model for the southern polar areas of the Moon (Esquivel Lizondo, 2024) and the installation of an array of VLF antennas in its PSRs to detect the first radiation waves coming from the dawn of the universe (Esquivel Lizondo, 2023). At last, Miguel Torres Torriti showed interesting R&D robotic technologies under study in the Robotics and Automation Laboratory (RAL) at PUC that optimize tire-terrain interactions (Prado et al., 2020) and visual grape bunch detection in vineyards (Pérez-Zavala et al., 2018) that can be useful in agricultural activities on a future lunar base. However, this is largely a photograph of the current situation, the point is how the future generations who are now students are going to develop it in the next years. As it will be shown in the next sections, there is a general interest of the most talented students towards space science but they are a small fraction of the total (2–3 boys and girls out of a class of 15 students on average) and will need a basic support of funding and studies to build up an expertise that can play a significant role in the Artemis Accords. It will take 5–10 years at least before they can have enough experience to start their careers as space scientists, technicians, and possibly entrepreneurs. One of the most problematic caveats is that Chile has a well-established astronomical research thanks to the various telescopes built on its territory but it is still under development, albeit with good quality, in terms of Solar System science.

6.4 Collaboration Patterns and Institutional Representation

A study by researchers at the University of Atacama revealed a concerning pattern in research collaborations related to astrobiology and planetary science in the region. Despite the immense scientific value of sites like the Atacama Desert, approximately 60% of publications based on data from the Puna, Altiplano, or Atacama Desert regions with objectives related to planetary science or astrobiology did not include any local institutional partners from Argentina, Bolivia, Chile, or Peru (Tavernier et al., 2023) (Fig. 6.4).

This pattern raises important questions about equitable scientific collaboration and knowledge production. The study suggests that Latin American planetary science would benefit from:

1. Strategic structuring and networking at national and continental scales
2. Development of a comprehensive road map for research, development, and innovation
3. Implementation of policies to protect these exceptional natural heritage sites while facilitating scientific research (Tavernier et al., 2023)

The authors note several examples of successful international collaborations in related fields, including meteorite studies, terrestrial analogues, space exploration in Chile and astrobiology initiatives in Mexico (Tavernier et al., 2023). These examples could serve as models for more inclusive collaboration frameworks in the future.

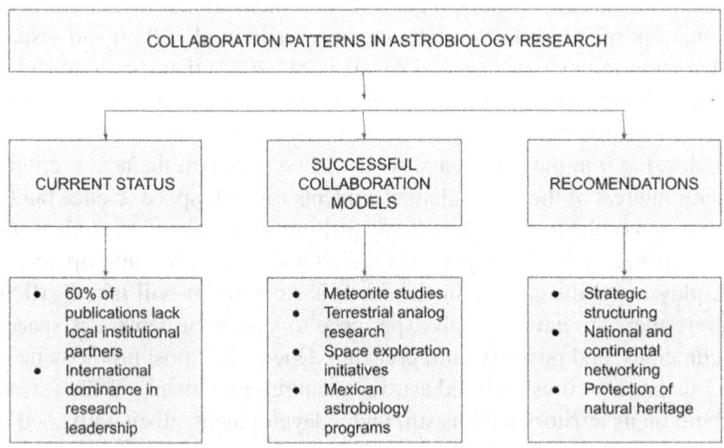

Fig. 6.4 Current Collaboration Patterns in Chilean Astrobiology Research

6.5 Education

An important engine for research in Planetary Sciences are the various doctorate (Ph.D.) programs in Astronomy, Astrophysics, Astronomy and Planetary Sciences, and even in Physics or Sciences with mention in Astronomy or Geology that some Chilean universities are successfully carrying out with the supervision of the national accreditation system (Comisión Nacional de Acreditación—CNA in Spanish). The accreditation system of the CNA ensures a continuous improvement of the quality of Ph.D. programs both in public and private universities of Chile (Espinoza & Eduardo González, 2013). For example, UAB offers a doctorate in Astrophysics[8] as well as PUC,[9] Universidad de Valparaiso (UV),[10] and UDP.[11] Other universities like UCH offer a doctorate in Sciences with mention in Astronomy,[12] Universidad La Serena (ULS) a doctorate in Astronomy,[13] UCN a doctorate in Sciences with Mention in Geology,[14] and UDA, a unique case in Chile and probably in the world, a doctorate in Astronomy and Planetary Sciences.[15] UDA is also planning a career in Sciences for bachelor students (Licenciatura in Spanish) with various mentions, including Astronomy, which lists elective courses in Planetary Sciences and it is scheduled to start in 2026. This career is designed to ensure a continuous flow of students for the Ph.D. in Astronomy and Planetary Sciences within the same university. In the case of UCH, Astronomy and Astrophysics courses are explicitly mentioned,[16] and elective courses of Planetary Sciences are available for postgraduate (and undergraduate) students. Other universities follow the same type of career in Science or Physics with mention in Astronomy or Astrophysics. However, the majority does not have, or it is not clear if they have, elective courses related to Planetary Sciences. UDP mentions courses in Solar System and Exoplanets. USM mentions a course of Interstellar Medium and Planetary Systems, something that looks more related to Exoplanets rather than our Solar System. In the case of UCH, Astronomy and Astrophysics courses are explicitly mentioned, and elective.[17] The same applies for Universidad de Concepcion (UDEC), PUC, UV, and almost all the other universities that carry out research about Astronomy in Chile.[18] Such a situation may appear disappointing and even

---

[8] https://investigacion.unab.cl/doctorados-eng/astrophysics/

[9] https://astro.uc.cl/en/doctorado-en-astrofisica/

[10] https://ifa.uv.cl/programa-academico/doctorado-enastrofisica/

[11] https://astronomia.udp.cl/en/teaching/phd-astrophysics/

[12] https://uchile.cl/postgrados/97546/astronomy-

[13] https://vipuls.userena.digital/es/postgrado/programas/doctorado/astronomia/

[14] https://ficg.ucn.cl/geodoctorado/

[15] http://inct.uda.cl/

[16] https://ingenieria.uchile.cl/carreras/4980/astronomia

[17] https://ingenieria.uchile.cl/carreras/4980/astronomia

[18] https://sochias.cl/astronomia-en-chile/universidades/,   https://www.astro.uantof.cl/educacion/magister-en-astronomia/

troublesome if Chile aims at fulfilling or simply contributing to the objectives stated in the Artemis Accords. However, the public outreach that will be discussed in the next subchapter and the potential for R&D shown at LAAP24 show encouraging signs that the current situation may change in the (hopefully near) future. Of particular note is the UA's Masters in Astronomy, which offers courses and thesis research in the field of astrobiology (see footnote 20).

6.6 Public Outreach

As already mentioned in the previous subchapter, some good news come from the situation of the public outreach in Chile. Last year 2024, the Chilean government, in collaboration with the Chilean Society of Astronomy (SOCHIAS in Spanish) and with the European Southern Observatory (ESO), celebrated the 10 years of the Astronomy Day in which actively patrocinated more than 100 activities all over the country.[19] Planetary news about Mars and the new Space Race to the Moon attract the interest of the general audience and many researchers offer seminars and dedicated events in Chilean schools and universities as part of their academic duties. The seminars dedicated to the general audience are generally offered in Spanish language but the seminars dedicated to the university students are more and more in English language, in order to stimulate the learning of the language, seminars for doctoral students are exclusively in English. This approach seems working as the number of students who are improving their language is increasing. Even those students who do not manage well the English language yet are still perfectly aware of the importance of this language in science and thus are doing admirable learning efforts through courses offered by the universities and by other Chilean private institutions. As a result, the number of students approaching to Planetary Sciences is increasing as well as the number of students willing to write a thesis on a Planetary subject related to the Moon or Mars. This is a good example on how a successful public outreach may help the education sector with fresh new entries of motivated students. Two or three talented students per course, on average, show a great potential for doing good in Planetary Science research.

6.7 Potential for Near-Future Investigations in Planetary Sciences

The high number of universities in Chile, coupled to the quality system of the CNA, the presence of natural laboratories, on this regard the Atacama Desert is a great case of study, and the high number of telescopes is increasing the potential

---

[19] https://www.minciencia.gob.cl/noticias/minciencia-celebra-el-dia-de-la-astronomia-2024-con-mas-de-100-actividades-en-el-pais/

for future investigation in the fields of Astrobiology and Planetary Sciences. The wealth of existing telescopes in Chile allows the search for potential biosignatures in exoplanets (Schwieterman et al., 2018). A new rising line of research is related to the multidisciplinary approach towards the spectroscopic analysis of existing meteorite collections and planetary simulant from Atacama Desert to compare to spectra acquired by the James Webb Space Telescope (JWST) on protoplanetary disks located in the Orion constellation (Batalla Falcón, 2020; Batalla-Falcon et al., 2025). Another rising line of research is related to experiments of growth of bacteria in the CAD-1 planetary simulant from Atacama Desert (San Martin et al., 2025). However, the most promising perspective for future investigations, including technological and economic development, comes from the participation of Chile to the Artemis Accords.[20] As a result, the organization of the first workshop in Santiago dedicated to the Artemis Project was aimed at building a national and international scientific ecosystem around existing expertise in Chile including all the fields of the human knowledge and Planetary Sciences working as an aggregating engine. Planetary Sciences and Planetary Geology, in particular, may certainly play a pivotal role where all the other fields can build up their strategies for the development of the future lunar base. The location and the mineral resources available in the lunar terrains are fundamental for the correct development of the future lunar base (Zhang et al., 2023). On this regard, existing Chilean expertise about resilient plants from Atacama Desert for oxygen and food production (Eshel et al., 2021) and more efficient bioleaching processes (Véliz et al., 2022) would be essential for the development of the future base in the environments of the Moon and Mars. Although the lunar poles are characterized by a mitigated environment, if compared to the extreme variations of temperature found at other latitudes, their cratered terrains still offer technological challenging conditions for the deployment of a surface temporary base aimed at the extraction of the resources and the excavation of the permanent base. An underground lunar base is necessary to avoid both risk of radiation and impact of micrometeorites (Haeuplik-Meusburger & Bannova, 2023). Also important, an underground lunar base would offer direct access to underground mineral and water resources that would be difficult to reach from the surface (Siekmeier & Podnieks, 1993). The permanently shadowed regions (PSRs) inside the craters offer a dark and very cold environment useful to run spectrographs without cryogenic cooling. This would be a much cheaper and significant advantage compared to cryogenically cooled spectrographs in the search for potential biosignatures in exoEarths (Rauscher et al., 2016). The lack of an atmosphere would make adaptive optics unnecessary for ground-based telescopes on the Moon.

---

[20] https://www.nasa.gov/news-release/nasa-welcomes-chile-as-newest-artemis-accords-signatory/#:~:text=The%20Republic%20of%20Chile%20is%20the%2047th%20country,cooperation%20among%20nations%20participating%20in%20NASA%E2%80%99s%20Artemis%20program

6.8 Future Directions and Research Prospects in Astrobiology

Based on current research trajectories, several promising directions for astrobiology research in Chile can be identified:

1. *Enhanced Local Participation:* Addressing the underrepresentation of local institutions could lead to more comprehensive and contextually informed research outcomes. This would involve developing strategic networking frameworks at national and continental levels (Tavernier et al., 2023).
2. *Advanced Biosignature Detection Technologies:* Building successful applications of techniques like Raman spectroscopy, future research could focus on increasing the sensitivity and specificity of biosignature detection methodologies (Veneranda et al., 2021; Shen et al., 2022; Abrahamsson & Kanik, 2022).
3. *Expanded Exploration of Microbial Diversity:* The identification of complex microbial communities in the Atacama suggests opportunities for further characterisation of extremophile adaptations, potentially informing the search for life in extraterrestrial environments (Aerts et al., 2020).
4. *Integrated Multi-Analytic Approaches:* Combining complementary analytical techniques (spectroscopic, molecular, and microbiological) could enhance detection capabilities and provide a more comprehensive characterisation of potential biosignatures.
5. *Cross-Disciplinary Integration:* Strengthening connections between astrobiology and related fields such as astronomy, geology, and atmospheric science could yield new insights into habitability conditions relevant to both terrestrial extreme environments and potential extraterrestrial habitats.

Chile's contribution to astrobiology research centres primarily on the exceptional Mars-analogue environments in the Atacama Desert. Major research initiatives like ARADS and ExoFiT have demonstrated the value of these sites for testing technologies and methodologies destined for deployment on Mars, particularly in sample collection, contamination control, and biosignature detection.

Research in the region has revealed that even in extreme environments characterised by high UV radiation, extreme aridity, and salt-dominated mineralogy, complex microbial communities can persist—findings that inform our understanding of the potential for life in similar extraterrestrial environments. The successful detection of organic functional groups using Raman spectroscopy demonstrates the feasibility of identifying potential biosignatures in Mars-like conditions.

However, the underrepresentation of local institutions in international research collaborations represents a significant challenge that could be addressed through strategic structuring and networking at national and continental scales. Enhancing local participation would promote more equitable knowledge production and enrich research outcomes by incorporating local expertise and contextual understanding.

As astrobiology continues to evolve as a discipline, Chile's unique environmental assets position it to make increasingly significant contributions to our understanding of life's potential beyond Earth, particularly through advanced technology development, refined biosignature detection methodologies, and expanded exploration of extremophile adaptations in Mars-analogue environments.

Declarations

Competing Interests The authors declare no competing interests.

References

Abrahamsson, V., & Kanik, I. (2022). In situ organic biosignature detection techniques for space applications. *Frontiers in Astronomy and Space Sciences, 9*, 959670. https://doi.org/10.3389/fspas.2022.959670

Aerts, J. W., Riedo, A., Melton, D. J., Martini, S., Flahaut, J., Meierhenrich, U. J., Meinert, C., Myrgorodska, I., Lindner, R., & Ehrenfreund, P. (2020). Biosignature analysis of Mars soil analogs from the Atacama Desert: Challenges and implications for future missions to Mars. *Astrobiology, 20*(6), 766–784. https://doi.org/10.1089/ast.2019.2063

Batalla Falcón, G. Á. (2020). *Estudio del espectro infrarrojo de meteoritos y aplicación en observaciones astronómicas.* Undergraduate thesis, Universidad Diego Portales.

Batalla-Falcon, G. A., Cieza, L. A., Lavin, R., Valenzuela, M., Morlok, A., Chavan, P., Farias, C., Leone, G., & Moncada, D. (2025). Mid-infrared absorption spectra and mass absorption coefficients for 23 chondrites—Dependence on composition and grain size. *Astronomy & Astrophysics, 696*, A66. https://doi.org/10.1051/0004-6361/202452540

Bonaccorsi, R., Glass, B., Moreno-Paz, M., García-Villadangos, M., Warren-Rhodes, K., Parro, V., Manchado, J. M., Wilhelm, M. B., & McKay, C. P. (2023). In situ real-time monitoring for aseptic drilling: Lessons learned from the Atacama rover astrobiology drilling studies contamination control strategy and implementation and application to the icebreaker mars life detection mission. *Astrobiology, 23*(12), 1303–1336. https://doi.org/10.1089/ast.2022.0133

Caqueo, N., Olave, D. R., Palharini, R. C., Gaglio, E., Palharini, R. S. A., & Savino, R. (2023). Inflatable aerodynamic decelerator for CubeSat reentry and recovery: IAD geometrical effects on the flowfield structure. *Aerospace Science and Technology, 141*, 108571. https://doi.org/10.1016/j.ast.2023.108571

Caqueo, N., Palharini, R. C., Palharini, R. S. A., Gaglio, E., & Savino, R. (2024). Inflatable aerodynamic decelerators for CubeSat reentry and recovery: Surface properties. *Aerospace Science and Technology, 149*, 109151. https://doi.org/10.1016/j.ast.2024.109151

Casassus, S., Wright, C. M., Marino, S., Maddison, S. T., Wootten, A., Roman, P., Pérez, S., Pinilla, P., Wyatt, M., & Moral, V. (2015). A compact concentration of large grains in the HD 142527 protoplanetary dust trap. *The Astrophysical Journal, 812*(2), 126. https://doi.org/10.1088/0004-637X/812/2/126

Cieza, L., Williams, J., Kourkchi, E., Andrews, S., Casassus, S., Graves, S., & Schreiber, M. R. (2015). A SCUBA-2 850-μm survey of protoplanetary discs in the IC 348 cluster. *Monthly Notices of the Royal Astronomical Society, 454*(2), 1909–1920. https://doi.org/10.1093/mnras/stv2044

Cocca y Esquivel, A. A. M. (2024). The moon agreement and the Artemis accords: Is Latin America at a crossroads? In *Space fostering Latin American societies: Developing the Latin American continent through space, Part 5* (pp. 1–15). Springer Nature Switzerland. https://doi.org/10.1007/978-3-031-40415-3_1

Cortés-Zuleta, P., Rojo, P., Wang, S., Hinse, T. C., Hoyer, S., Sanhueza, B., Correa-Amaro, P., & Albornoz, J. (2020). TraMoS-V. Updated ephemeris and multi-epoch monitoring of the hot

Jupiters WASP-18Ab, WASP-19b, and WASP-77Ab. *Astronomy & Astrophysics, 636,* A98. https://doi.org/10.1051/0004-6361/201936279

de Pater, I., Davies, A. G., McGregor, A., Trujillo, C., Ádámkovics, M., Veeder, G. J., Matson, D. L., Leone, G., & the Gemini Io Team. (2014). Global near-IR maps from Gemini-N and Keck in 2010, with a special focus on Janus Patera and Kanehekili Fluctus. *Icarus, 242,* 379–395. https://doi.org/10.1016/j.icarus.2014.06.019

Eshel, G., Araus, V., Undurraga, S., Soto, D. C., Moraga, C., Montecinos, A., Moyano, T., Maldonado, J., Díaz, F. P., Varala, K., Nelson, C. W., Contreras-López, O., Pal-Gabor, H., Kraiser, T., Carrasco-Puga, G., Nilo-Poyanco, R., Zegar, C. M., Orellana, A., Montecino, M., Maass, A., Allende, M. L., DeSalle, R., Stevenson, D. W., González, M., Latorre, C., Coruzzi, G. M., & Gutiérrez, R. A. (2021). Plant ecological genomics at the limits of life in the Atacama Desert. *Proceedings of the National Academy of Sciences, 118*(46), e2101177118. https://doi. org/10.1073/pnas.2101177118

Espinoza, Ó., & Eduardo González, L. (2013). Accreditation in higher education in Chile: Results and consequences. *Quality Assurance in Education, 21*(1), 20–38. https://doi. org/10.1108/09684881311293043

Esquivel Lizondo, M. E. (2023). Proyecto VLF lunar. *Revista Española de Derecho Aeronáutico y Espacial, 3,* 591–606.

Esquivel Lizondo, M. E. (2024). Propuesta de gobernanza de las zonas lunares especialmente administradas en el polo sur lunar: Un modelo basado en el tratado antártico. *Revista Española de Derecho Aeronáutico y Espacial, 4,* 383–397.

Fishbaugh, K. E., Des Marais, D. J., Korablev, O., Lognonne, P., & Raulin, F. (2007). Introduction: A multidisciplinary approach to habitability. *Geology and Habitability of Terrestrial Planets,* 1–5. https://doi.org/10.1007/978-0-387-74288-5_1

Gattacceca, J., Valenzuela, M., Uehara, M., Jull, A. J. T., Giscard, M., Rochette, P., Braucher, R., Suavet, C., Gounelle, M., Morata, D., Munayco, P., Bourot-Denise, M., Bourlès, D., & Demory, F. (2011). The densest meteorite collection area in hot deserts: The San Juan meteorite field (Atacama Desert, Chile). *Meteoritics & Planetary Science, 46*(9), 1276–1287. https://doi. org/10.1111/j.1945-5100.2011.01229.x

Gerón-Piñón, G., Solana-González, P., Trigueros-Preciado, S., & Pérez-González, D. (2021). Management indicators: Their impact on Latin-American universities' accreditation. *Quality in Higher Education, 27*(2), 184–205. https://doi.org/10.1080/13538322.2021.1890318

Glass, B., Bergman, D., Parro, V., Kobayashi, L., Stoker, C., Quinn, R., Davila, A., Willis, P., Brinckerhoff, W., Warren-Rhodes, K., Wilhelm, M. B., Caceres, L., DiRuggiero, J., Zacny, K., Moreno-Paz, M., Dave, A., Seitz, S., Grubisic, A., Castillo, M., Bonaccorsi, R., & ARADS Team. (2023). The Atacama Rover Astrobiology Drilling Studies (ARADS) Project. *Astrobiology, 23*(12), 1245–1258. https://doi.org/10.1089/ast.2022.0126

Guzmán, V. V., Bergner, J. B., Law, C. J., Öberg, K. I., Walsh, C., Cataldi, G., Aikawa, Y., Bergin, E. A., Czekala, I., Huang, J., Andrews, S. M., Loomis, R. A., Zhang, K., Le Gal, R., Alarcón, F., Ilee, J. D., Teague, R., Cleeves, L. I., Wilner, D. J., Long, F., Schwarz, K. R., Bosman, A. D., Pérez, L. M., Ménard, F., & Liu, Y. (2021). Molecules with ALMA at planet-forming scales (MAPS). VI. Distribution of the small organics HCN, C2H, and H2CO. *The Astrophysical Journal Supplement Series, 257*(1), 6. https://doi.org/10.3847/1538-4365/ac1440

Haeuplik-Meusburger, S., & Bannova, O. (2023). Reflections on early lunar base design–From sketch to the first moon landing. *Acta Astronautica, 202,* 729–741. https://doi.org/10.1016/j. actaastro.2022.09.021

Hoyer, S., Rojo, P., López-Morales, M., Díaz, R. F., Chambers, J., & Minniti, D. (2011). Five new transit epochs of the exoplanet OGLE-TR-111b. *The Astrophysical Journal, 733*(1), 53. https:// doi.org/10.1088/0004-637X/733/1/53

Hutzler, A., Gattacceca, J., Rochette, P., Braucher, R., Carro, B., Christensen, E. J., Cournede, C., Gounelle, M., Laridhi Ouazaa, N., Martinez, R., Valenzuela, M., Warner, M., & Bourles, D. (2016). Description of a very dense meteorite collection area in western Atacama: Insight into the long-term composition of the meteorite flux to Earth. *Meteoritics & Planetary Science, 51*(3), 468–482. https://doi.org/10.1111/maps.12607

Ivanov, V. D., Beamín, J. C., Cáceres, C., & Minniti, D. (2020). A qualitative classification of extraterrestrial civilizations. *Astronomy & Astrophysics, 639*, A94.

Jenkins, J. S., Jones, H. R. A., Tuomi, M., Díaz, M., Cordero, J. P., Aguayo, A., Pantoja, B., Arriagada, P., Mahu, R., Brahm, R., Rojo, P., Soto, M. G., Ivanyuk, O., Becerra Yoma, N., Day-Jones, A. C., Ruiz, M. T., Pavlenko, Y. V., Barnes, J. R., Murgas, F., Pinfield, D. J., Jones, M. I., López-Morales, M., Shectman, S., Butler, R. P., & Minniti, D. (2017). New planetary systems from the Calan–Hertfordshire extrasolar planet search. *Monthly Notices of the Royal Astronomical Society, 466*(1), 443–473. https://doi.org/10.1093/mnras/stw2811

Jenkins, J. S., Harrington, J., Challener, R. C., Kurtovic, N. T., Ramirez, R., Peña, J., McIntyre, K. J., Himes, M. D., Rodríguez, E., Anglada-Escudé, G., Dreizler, S., Ofir, A., Peña Rojas, P. A., Ribas, I., Rojo, P., Kipping, D., Butler, R. P., Amado, P. J., Rodríguez-López, C., Kempton, E. M.-R., Pallé, E., & Murgas, F. (2019). Proxima Centauri b is not a transiting exoplanet. *Monthly Notices of the Royal Astronomical Society, 487*(1), 268–274. https://doi.org/10.1093/mnras/stz1268

Jenkins, J. S., Díaz, M. R., Kurtovic, N. T., Espinoza, N., Vines, J. I., Peña Rojas, P. A., Brahm, R., Torres, P., Cortés-Zuleta, P., Soto, M. G., Lopez, E. D., King, G. W., Wheatley, P. J., Winn, J. N., Ciardi, D. R., Ricker, G., Vanderspek, R., Latham, D. W., Seager, S., Jenkins, J. M., Beichman, C. A., Bieryla, A., Burke, C. J., Christiansen, J. L., Henze, C. E., Klaus, T. C., McCauliff, S., Mori, M., Narita, N., Nishiumi, T., Tamura, M., Pitogo de Leon, J., Quinn, S. N., Villaseñor, J. N., Vezie, M., Lissauer, J. J., Collins, K. A., Collins, K. I., Isopi, G., Mallia, F., Ercolino, A., Petrovich, C., Jordán, A., Acton, J. S., Armstrong, D. J., Bayliss, D., Bouchy, F., Belardi, C., Bryant, E. M., Burleigh, M. R., Cabrera, J., Casewell, S. L., Chaushev, A., Cooke, B. F., Eigmüller, P., Erikson, A., Foxell, E., Gänsicke, B. T., Gill, S., Gillen, E., Günther, M. N., Goad, M. R., Hooton, M. J., Jackman, J. A. G., Louden, T., McCormac, J., Moyano, M., Nielsen, L. D., Pollacco, D., Queloz, D., Rauer, H., Raynard, L., Smith, A. M. S., Tilbrook, R. H., Titz-Weider, R., Turner, O., Udry, S., Walker, S. R., Watson, C. A., West, R. G., Pallé, E., Ziegler, C., Law, N., & W., A. (2020). An ultrahot Neptune in the Neptune desert. *Nature Astronomy, 4*(12), 1148–1157. https://doi.org/10.1038/s41550-020-1142-z

Jones, M. I., Jenkins, J. S., Bluhm, P., Rojo, P., & Melo, C. H. F. (2014). The properties of planets around giant stars. *Astronomy & Astrophysics, 566*, A113. https://doi.org/10.1051/0004-6361/201323345

Kraus, A. L., Ireland, M. J., Cieza, L. A., Hinkley, S., Dupuy, T. J., Bowler, B. P., & Liu, M. C. (2013). Three wide planetary-mass companions to FW Tau, ROXs 12, and ROXs 42B. *The Astrophysical Journal, 781*(1), 20. https://doi.org/10.1088/0004-637X/781/1/20

Krauskopf, E. (2021). Article processing charge expenditure in Chile: The current situation. *Learned Publishing, 34*(4), 637–646. https://doi.org/10.1002/leap.1413

Krone, L. V., Hampl, F. J., Schwerdhelm, C., Bryce, C., Ganzert, L., Kitte, A., Übernickel, K., Dielforder, A., Aldaz, S., Oses-Pedraza, R., Perez, J. P. H., Sanchez-Alfaro, P., Wagner, D., Weckmann, U., & von Blanckenburg, F. (2021). Deep weathering in the semi-arid Coastal Cordillera, Chile. *Scientific Reports, 11*(1), 13057. https://doi.org/10.1038/s41598-021-90267-7

Lamarra, N. F. (2003). Higher education, quality evaluation and accreditation in Latin America and MERCOSUR. *European Journal of Education, 38*(3), 253–269.

Leone, G. (2014). A network of lava tubes as the origin of Labyrinthus Noctis and Valles Marineris on Mars. *Journal of Volcanology and Geothermal Research, 277*, 1–8. https://doi.org/10.1016/j.jvolgeores.2014.01.011

Leone, G. (2016). Alignments of volcanic features in the southern hemisphere of Mars produced by migrating mantle plumes. *Journal of Volcanology and Geothermal Research, 309*, 78–95. https://doi.org/10.1016/j.jvolgeores.2015.10.028

Leone, G. (2017). Mangala Valles, Mars: A reassessment of formation processes based on a new geomorphological and stratigraphic analysis of the geological units. *Journal of Volcanology and Geothermal Research, 337*, 62–80. https://doi.org/10.1016/j.jvolgeores.2017.03.011

Leone, G., Tackley, P. J., Gerya, T. V., May, D. A., & Zhu, G. (2014). Three-dimensional simulations of the southern polar giant impact hypothesis for the origin of the Martian dichotomy. *Geophysical Research Letters, 41*(24), 8736–8743. https://doi.org/10.1002/2014GL062261

Leone, G., Ahrens, C., Korteniemi, J., Gasparri, D., Kereszturi, A., Martynov, A., et al. (2023). Sverdrup-Henson crater: A candidate location for the first lunar South Pole settlement. *Iscience, 26*(10).

Lukić, D. V. (2017). Search for possible exomoons with the FAST telescope. *Research in Astronomy and Astrophysics, 17*(12), 121. https://doi.org/10.1088/1674-4527/17/12/121

Molina-Montenegro, M. A., Galleguillos, C., Oses, R., Acuña-Rodríguez, I. S., Lavín, P., Gallardo-Cerda, J., Torres-Díaz, C., Diez, B., Pizarro, G. E., & Atala, C. (2016a). Adaptive phenotypic plasticity and competitive ability deployed under a climate change scenario may promote the invasion of Poa annua in Antarctica. *Biological Invasions, 18*, 603–618. https://doi.org/10.1007/s10530-015-1033-x

Molina-Montenegro, M. A., Oses, R., Atala, C., Torres-Díaz, C., Bolados, G., & León-Lobos, P. (2016b). Nurse effect and soil microorganisms are key to improve the establishment of native plants in a semiarid community. *Journal of Arid Environments, 126*, 54–61. https://doi.org/10.1016/j.jaridenv.2015.10.016

Montesinos, M., Cuadra, J., Perez, S., Baruteau, C., & Casassus, S. (2015). Protoplanetary disks including radiative feedback from accreting planets. *The Astrophysical Journal, 806*(2), 253. https://doi.org/10.1088/0004-637X/806/2/253

Mutis, T. C., Jara, N. C., Sepúlveda, D. V., Stickar, J. P., & Palharini, R. C. (2024). Numerical investigation of micropropulsion systems for CubeSats: Gas species and geometrical effects on nozzle performance. *Aerospace Science and Technology, 155*, 109625. https://doi.org/10.1016/j.ast.2024.109625

Oses Pedraza, R., Mitzscherling, J., Lipus, D., Wagner, D., & Bartholomäus, A. (2023). Genome Sequence of Arthrobacter sp. Strain ATA002, a Seed Endophytic Bacterium from the Atacama Desert. *Microbiology Resource Announcements, 12*(5), e00027–e00023. https://doi.org/10.1128/mra.00027-23

Parro, V., de Diego-Castilla, G., Moreno-Paz, M., Blanco, Y., Cruz-Gil, P., Rodríguez-Manfredi, J. A., Fernández-Remolar, D., Gómez, F., Gómez, M. J., Rivas, L. A., Demergasso, C., Echeverría, A., Urtuvia, V. N., Ruiz-Bermejo, M., García-Villadangos, M., Postigo, M., Sánchez-Román, M., Chong-Díaz, G., & Gómez-Elvira, J. (2011). A microbial oasis in the hypersaline Atacama subsurface discovered by a life detector chip: Implications for the search for life on Mars. *Astrobiology, 11*(10), 969–996. https://doi.org/10.1089/ast.2011.0654

Parro, V., Blanco, Y., Puente-Sánchez, F., Rivas, L. A., Moreno-Paz, M., Echeverría, A., Chong-Díaz, G., Demergasso, C., & Cabrol, N. A. (2018). Biomarkers and metabolic patterns in the sediments of evolving glacial lakes as a proxy for planetary lake exploration. *Astrobiology, 18*(5), 586–606. https://doi.org/10.1089/ast.2015.1342

Pérez, S., Marino, S., Casassus, S., Baruteau, C., Zurlo, A., Flores, C., & Chauvin, G. (2019). Upper limits on protolunar disc masses using ALMA observations of directly imaged exoplanets. *Monthly Notices of the Royal Astronomical Society, 488*(1), 1005–1011. https://doi.org/10.1093/mnras/stz1775

Pérez-Zavala, R., Torres-Torriti, M., Cheein, F. A., & Troni, G. (2018). A pattern recognition strategy for visual grape bunch detection in vineyards. *Computers and Electronics in Agriculture, 151*, 136–149. https://doi.org/10.1016/j.compag.2018.05.019

Petrovich, C., Muñoz, D. J., Kratter, K. M., & Malhotra, R. (2020). A disk-driven resonance as the origin of high inclinations of close-in planets. *The Astrophysical Journal Letters, 902*(1), L5. https://doi.org/10.3847/2041-8213/abb952

Pinto, G. A., Tavernier, A., Gattacceca, J., Corgne, A., Valenzuela, M., Luais, B., Flores, L., Olivares, F., & Marrocchi, Y. (2024). Dense collection areas and terrestrial alteration of meteorites in the Atacama Desert. *Meteoritics & Planetary Science, 59*(2), 351–367. https://doi.org/10.1111/maps.14125

Prado, Á. J., Torres-Torriti, M., Yuz, J., & Cheein, F. A. (2020). Tube-based nonlinear model predictive control for autonomous skid-steer mobile robots with tire–terrain interactions. *Control Engineering Practice, 101*, 104451. https://doi.org/10.1016/j.conengprac.2020.104451

Rauscher, B. J., Canavan, E. R., Moseley, S. H., Sadleir, J. E., & Stevenson, T. (2016). Detectors and cooling technology for direct spectroscopic biosignature characterization. *Journal of Astronomical Telescopes, Instruments, and Systems, 2*(4), 041212–041212. https://doi.org/10.1117/1.JATIS.2.4.041212

San Martin, J., Leone, G., Riveros-Jensen, K., Alam, M. A., Cabrera, R., San Martin, D., Oses, R., Blamey, J. M., Demergasso, C., Abrevaya, X. C., Guiliani, N., Britt, D. T., Liu, Y., Silva, W., Slumba, K., Tovar, D., Leal, M. A., & de Pablo, M. A. (2025). Prospecting the first Chilean Martian simulants from the Atacama Desert for ISRU and potential applications. *Icarus, 429,* 116403. https://doi.org/10.1016/j.icarus.2024.116403

Santos, C. M. D., Alabi, L. P., Friaça, A. C., & Galante, D. (2016). On the parallels between cosmology and astrobiology: A transdisciplinary approach to the search for extraterrestrial life. *International Journal of Astrobiology, 15*(4), 251–260. https://doi.org/10.1017/S1473550416000094

Schwieterman, E. W., Kiang, N. Y., Parenteau, M. N., Harman, C. E., DasSarma, S., Fisher, T. M., et al. (2018). Exoplanet biosignatures: A review of remotely detectable signs of life. *Astrobiology, 18*(6), 663–708. https://doi.org/10.1089/ast.2017.1729

Shen, J., Chen, Y., Sun, Y., Liu, L., Pan, Y., & Lin, W. (2022). Detection of biosignatures in Terrestrial analogs of Martian regions: Strategical and technical assessments. *Earth and Planetary Physics, 6*(5), 431–450. https://doi.org/10.26464/epp2022042.pdf

Siekmeier, J., & Podnieks, E. (1993). Underground lunar base and mine-A desired combination. In *Space programs and technologies conference and exhibit.* https://doi.org/10.2514/6.1993-4135.

Tavernier, A., Pinto, G. A., Valenzuela, M., Garcia, A., Ulloa, C., Oses, R., & Foing, B. H. (2023). Trends in planetary science research in the Puna and Atacama Desert regions: Underrepresentation of local scientific institutions? *Meteoritics & Planetary Science, 58*(4), 516–528. https://doi.org/10.1111/maps.13972

Torres-Díaz, C., Gallardo-Cerda, J., Lavin, P., Oses, R., Carrasco-Urra, F., Atala, C., Acuña-Rodríguez, I. S., Convey, P., & Molina-Montenegro, M. A. (2016). Biological interactions and simulated climate change modulates the ecophysiological performance of Colobanthus quitensis in the Antarctic ecosystem. *PLoS One, 11*(10), e0164844. https://doi.org/10.1371/journal.pone.0164844

Tregloan-Reed, J., Southworth, J., Burgdorf, M., Calchi Novati, S., Dominik, M., Finet, F., Jørgensen, U. G., Maier, G., Mancini, L., Prof, S., Ricci, D., Snodgrass, C., Bozza, V., Browne, P., Dodds, P., Gerner, T., Harpsøe, K., Hinse, T. C., Hundertmark, M., Kains, N., Kerins, E., Liebig, C., Penny, M. T., Rahvar, S., Sahu, K., Scarpetta, G., Schäfer, S., Schönebeck, F., Skottfelt, J., & Surdej, J. (2015). Transits and starspots in the WASP-6 planetary system. *Monthly Notices of the Royal Astronomical Society, 450*(2), 1760–1769. https://doi.org/10.1093/mnras/stv730

Valenzuela, M., & Benado, J. (2018). Meteorites and craters found in Chile: A bridge to introduce the first attempt for geoheritage legal protection in the country. In R. Acevedo & J. Frías (Eds.), *Geoethics in Latin America* (The Latin American Studies Book Series) (pp. 103–115). Springer. https://doi.org/10.1007/978-3-319-75373-7_7

Véliz, D., & Celis, S. (2021). The development of the research capabilities of Chilean faculty. In T. Aarrevaara, M. Finkelstein, G. A. Jones, & J. Jung (Eds.), *Universities in the Knowledge Society: The Nexus of National Systems of Innovation and Higher Education* (pp. 339–355). Springer. https://doi.org/10.1007/978-3-030-76579-8_19

Véliz, M. V., Leiva, A. V., & Bellange, P. M. (2022). Copper Bioleaching Operations in Chile: Towards New Challenges. In *Biomining technologies: Extracting and recovering metals from ores and wastes* (p. 163).

Veneranda, M., Lopez-Reyes, G., Saiz, J., Manrique-Martinez, J. A., Sanz-Arranz, A., Medina, J., Moral, A., Seoane, L., Ibarmia, S., & Rull, F. (2021). ExoFiT trial at the Atacama Desert (Chile): Raman detection of biomarkers by representative prototypes of the ExoMars/Raman Laser Spectrometer. *Scientific Reports, 11*(1), 1461. https://doi.org/10.1038/s41598-021-81014-z

Wei, J., Wang, A., Lambert, J. L., Wettergreen, D., Cabrol, N., Warren-Rhodes, K., & Zacny, K. (2015). Autonomous soil analysis by the Mars Micro-beam Raman Spectrometer (MMRS) on-board a rover in the Atacama Desert: A terrestrial test for planetary exploration. *Journal of Raman Spectroscopy, 46*(10), 810–821. https://doi.org/10.1002/jrs.4656

Zhang, P., Dai, W., Niu, R., Zhang, G., Liu, G., Liu, X., et al. (2023). Overview of the lunar in situ resource utilization techniques for future lunar missions. *Space: Science & Technology, 3*, 0037. https://doi.org/10.34133/space.0037

Chapter 7
Astrobiology and Planetary Sciences in Colombia

David Tovar, María Angélica Leal, Jimena Sánchez, Rosa Reyes,
Trinidad Ceferino, Argenis Bonilla, Nadejda Tchegliakova,
María Camila Orozco, Luz Marina Melgarejo, Liliana Figueroa,
Rosa Acevedo-Barrios, Walter Ocampo, Carlos Molina,
Alejandro Guerrero-Caicedo, and Jonathan Pelegrin

D. Tovar (✉)
Astrobiology and Planetary Sciences Research Group GCPA, Universidad Nacional de
Colombia & Corporación Científica Laguna, Bogotá, Colombia

Geosciences Department, Faculty of Sciences, Universidad Nacional de Colombia,
Bogotá, Colombia

Unit of Geology, Universidad de Alcalá, Alcalá de Henares, Spain
e-mail: dftovarr@unal.edu.co

M. A. Leal
Astrobiology and Planetary Sciences Research Group GCPA, Universidad Nacional de
Colombia & Corporación Científica Laguna, Bogotá, Colombia

Unit of Geology, Universidad de Alcalá, Alcalá de Henares, Spain

Biology Department, Faculty of Sciences, Universidad Nacional de Colombia,
Bogotá, Colombia

J. Sánchez
Astrobiology and Planetary Sciences Research Group GCPA, Universidad Nacional de
Colombia & Corporación Científica Laguna, Bogotá, Colombia

Biology Department, Faculty of Sciences, Universidad Nacional de Colombia,
Bogotá, Colombia

R. Reyes · T. Ceferino
Astrobiology and Planetary Sciences Research Group GCPA, Universidad Nacional de
Colombia & Corporación Científica Laguna, Bogotá, Colombia

A. Bonilla
Biology Department, Faculty of Sciences, Universidad Nacional de Colombia,
Bogotá, Colombia

Biology of Tropical Organisms Research Group, Universidad Nacional de Colombia,
Bogotá, Colombia

D. Tovar, M. A. Leal (eds.), *Astrobiology and Planetary Sciences in Latin
America*, https://doi.org/10.1007/978-3-032-01450-4_7

Abstract This chapter presents a historical and contemporary review of the development of astrobiology and planetary sciences in Colombia, highlighting their origins, progress, and future projections. Beginning with the emblematic fall of the Santa Rosa de Viterbo meteorite in 1810—an event that symbolizes the country's initial engagement with cosmic knowledge—the chapter traces the key milestones in research, education, outreach, and the formation of scientific communities. It examines the early initiatives in exobiology, the consolidation of research groups, the role of museums and planetariums, as well as strategies for the social appropriation of scientific knowledge. Furthermore, it explores major research

N. Tchegliakova
Astrobiology and Planetary Sciences Research Group GCPA, Universidad Nacional de Colombia & Corporación Científica Laguna, Bogotá, Colombia

Geosciences Department, Faculty of Sciences, Universidad Nacional de Colombia, Bogotá, Colombia

M. C. Orozco
Astrobiology and Planetary Sciences Research Group GCPA, Universidad Nacional de Colombia & Corporación Científica Laguna, Bogotá, Colombia

Faculty of Health Sciences, Fundación Universitaria San Martín, Bogotá, Colombia

L. M. Melgarejo
Astrobiology and Planetary Sciences Research Group GCPA, Universidad Nacional de Colombia & Corporación Científica Laguna, Bogotá, Colombia

Biology Department, Faculty of Sciences, Universidad Nacional de Colombia, Bogotá, Colombia

Stress Physiology and Biodiversity Group in Plants and Microorganisms, Universidad Nacional de Colombia, Bogotá, Colombia

L. Figueroa
Astrobiology and Planetary Sciences Research Group GCPA, Universidad Nacional de Colombia & Corporación Científica Laguna, Bogotá, Colombia

Environmental Engineering Program, Universidad El Bosque, Bogotá, Colombia

R. Acevedo-Barrios
Astrobiology and Planetary Sciences Research Group GCPA, Universidad Nacional de Colombia & Corporación Científica Laguna, Bogotá, Colombia

Grupo de Estudios Químicos y Biológicos, Dirección de Ciencias Básicas, Universidad Tecnológica de Bolívar, Cartagena de Indias, Colombia

W. Ocampo
Asociación de Estudios Astronómicos ACDA, Bogotá, Colombia

C. Molina
Astrobiology and Planetary Sciences Research Group GCPA, Universidad Nacional de Colombia & Corporación Científica Laguna, Bogotá, Colombia

Department of Science and Arts Education, Faculty of Education, Universidad de Antioquia, Medellín, Colombia

Department of Design, School of Arts and Humanities, Instituto Tecnológico Metropolitano, Medellín, Colombia lines in extreme environments and planetary analogs, alongside the development of educational programs and specialized scientific events. Finally, it outlines the perspectives that position Colombia as an emerging actor in the field of space sciences, leveraging its biodiversity, geodiversity, and the strengthening of interdisciplinary networks.

7.1 Introduction

1810 is remembered as the beginning of Colombia's independence process, a political and social milestone that profoundly transformed the destiny of the territory and its inhabitants (Pita Pico, 2021; Brungardt, 2000). However, that same year, an event of cosmic origin would silently and unexpectedly mark the course of Planetary Sciences in the country: the fall of the Santa Rosa de Viterbo meteorite, in the department of Boyacá. On the night of Holy Saturday, a shower of meteoritic masses culminated in the fall of a large metallic rock on the Tocavita hill, near the town of Santa Rosa de Viterbo. A young local resident, Cecilia Corredor, discovered a fragment of the celestial body, which was later moved to the town center (Rodríguez Prada, 2014). For years, the piece was used as an anvil by a local blacksmith, eventually becoming a symbol of the community and, ultimately, one of the foundational stones of scientific thought in the nascent Republic (Rodríguez Prada, 2014; Segura, 1995).

The so-called "aerolite of Santa Rosa de Viterbo," now classified as an anomalous metallic meteorite (Bsdok et al., 2020), is undoubtedly one of the most emblematic objects in the history of Planetary Sciences in Colombia. Its story—marked by neglect, wonder, looting, and eventual vindication—paradigmatically reflects the country's relationship with scientific knowledge. It was not until 1822—more than a decade after its fall—that the object began to be examined from a scientific perspective, thanks to the arrival of two young naturalists: Mariano Eduardo de Rivero y Ustariz from Peru, and Jean-Baptiste Boussingault from France. Both were

A. Guerrero-Caicedo
Semillero de investigación en Astroquímica y Astrobiología (AstroLUCA), Cali, Colombia

Faculty of Health Sciences, Universidad Libre, Cali, Colombia

Chemistry Department, Faculty of Natural and Exact Sciences, Universidad del Valle, Cali, Colombia

J. Pelegrin
Semillero de investigación en Astroquímica y Astrobiología (AstroLUCA), Cali, Colombia

Master's Program in Environmental Education and Sustainable Development, Facultad de Educación, Universidad Santiago de Cali, Cali, Colombia

Biology Department, Faculty of Natural and Exact Sciences, Universidad del Valle, Cali, Colombia

Biology Program, Department of Natural Sciences and Mathematics, Faculty of Engineering and Sciences, Pontificia Universidad Javeriana Cali, Cali, Colombia recruited in Paris by the diplomatic envoy Francisco Antonio Zea to join a scientific mission aimed at laying the foundations of natural knowledge in Colombia (Acevedo et al., 2014; Rodríguez Prada, 2010).

The contracts of Rivero and Boussingault outlined fundamental responsibilities: the establishment of a Museum of Natural Sciences, the design of a National School of Mines inspired by the Parisian model, and the systematic collection of geochemical, mineralogical, and astronomical observations (Ramírez, 1949, 1951; Ward, 1907). During their visit to Santa Rosa de Viterbo, these naturalists conducted a chemical analysis of the aerolite and communicated their findings to scientific institutions in Europe. Their observations were presented at the Academy of Sciences of the Institute of France by prominent figures such as Alexander von Humboldt and Georges Cuvier, and were soon published in scientific journals in Paris, London, and Leipzig (Ramírez, 1949).

In 1823, Rivero and Boussingault printed in Bogota: *Memoria sobre diferentes masas de hierro encontradas en la cordillera oriental de los Andes* ("Report on Various Masses of Iron Found in the Eastern Cordillera of the Andes"), in which they detailed their chemical analysis and described the aerolite, supported by lithographs produced in the most advanced printing workshop in the country at the time. This publication can be regarded as the first local geochemical study of a meteorite in Colombia and marked a transformation of the object—from an unknown natural mass to an item of scientific and cultural interest (Rodríguez Prada, 2014).

The symbolic and scientific value of the aerolite thus became embedded within a foundational moment for Colombian science. In 1823, with the institutional consolidation of the Republic under the vice presidency of Francisco de Paula Santander, Congress officially approved the establishment of the Museum of Natural Sciences— now the National Museum—and the School of Mines, both housed in the former Botanical House of José Celestino Mutis. The Museum was formally inaugurated in 1824, and among its first collections was a fragment of the iconic meteorite from Santa Rosa de Viterbo (Pulido Chaparro, 2015).

Subsequently, the aerolite became the object of desire for international collectors such as Henry A. Ward, who attempted to steal it in 1906 (Escallón, 2015). The incident triggered a national scandal that led to its recovery by order of President Rafael Reyes and its subsequent study at the National University of Colombia (Universidad Nacional de Colombia), marking a new chapter in the institutionalization of scientific knowledge concerning extraterrestrial objects in the country (Universidad Nacional de Colombia, n.d.). Years later, in 1949, Jesús Emilio Ramírez made the first Colombian contribution to the renowned journal *Contributions of the Meteoritical Society*, presenting the history and characterization of the Colombian aerolite (Ramírez, 1949). Finally, in 1992, the main fragment of the meteorite was reintegrated into the National Museum of Colombia (Museo Nacional de Colombia), where it is now preserved under catalog number 874 as part of the "Scientific Objects" collection (Paredes, 2016). Its placement at the center of the museum's main exhibition hall is no coincidence: it represents the beginning of the national museological narrative, the birth of a scientific gaze toward the cosmos, and a point of convergence between natural history, geology, astronomy, and the very construction of the Colombian state.

Fig. 7.1 Santa Rosa de Viterbo Meteorite. Fragment on display at the National University of Colombia. (Source: Captured and provided by Miguel Ángel de Pablo, 2025)

Just as the history of the Santa Rosa de Viterbo aerolite (Fig. 7.1) is rooted in that singular event which occurred in the very same year the Colombian nation was born, perhaps, as a country, we should have understood that cosmic message not only in the context of our independence, but also as a call to recognize that planetary sciences—particularly planetary geology and astrobiology—can occupy a central role in the advancement and development of the nation's space sciences. This is especially relevant considering that Colombia is not only one of the most biodiverse countries in the world, but also one of the most geodiverse (Tavera, 2015). Therefore, the present chapter seeks to highlight some of the progress Colombia has made in these fields.

7.2 First Steps

The term astrobiology was first introduced globally in 1953, when Soviet scientist Gavriil Adrianovich Tikhov published his book Астробиология (Astrobiology) (Tejfel, 2009). However, in the Americas, the term exobiology gained greater popularity, largely due to the Soviet origin of the term astrobiology and the prevailing geopolitical context of the time. It was not until 1964 that George Gaylord Simpson emphasized the lack of a clearly defined object of study if life was considered solely as extraterrestrial in origin, and if the field referred only to life beyond our planet (Morange, 2007). The widely disseminated term exobiology remained in use for many years, until 1998, when NASA established the program that would later become the NASA Astrobiology Institute (NAI) (Leal et al., 2015).

It is thus under the framework of the term exobiology that this scientific field finds its origins in Colombia. In 1986, a group of individuals deeply motivated by astronomy and its related sciences—driven by a desire to study and explore various topics while also engaging others and disseminating knowledge—founded the Colombian Association for Astronomical Studies (Asociación Colombiana de Estudios Astronómicos, ACDA). Since its inception, one of ACDA's primary mechanisms has been the formation of study groups in diverse areas, such as astronautics, astrophysics, cosmology, the Solar System, the history of astronomy, and, notably, astrobiology. By the end of 1986, members of ACDA regularly visited libraries to consult the journal *Origins of Life*, which served as the inspiration for the establishment of the Exobiology Commission in 1987. This was the association's first study group and, as far as is known, the first in Colombia to formally address this subject. The group was initially composed of two physicians, a biologist, a microbiologist, and a chemist, who began by studying the biological experiments conducted during the Viking mission to Mars—experiments that, even then, were the subject of considerable debate. Over time, the members of the study group began organizing public outreach lectures, which would eventually become one of ACDA's foundational pillars: sharing and outreaching high-quality scientific knowledge for the broader public.

These activities developed by ACDA led to its recognition in 1992 by the Administrative Department of Science, Technology, and Innovation, Colciencias, through its support program for scientific associations. As a result, ACDA was granted its own headquarters in the Camilo Torres Building at the Universidad Nacional de Colombia (National University of Colombia), where the association remained for several years (ACDA, 2022). These efforts began to generate interest and motivation among undergraduate students at the National University of Colombia, leading to the formation in 2004 of the group "Vida sin Fronteras" ("Life Without Borders")—later renamed the Astrobiology Club at the Universidad Nacional de Colombia (UNASB)—which focused on outreach activities within the university community and in public spaces such as city libraries. Outreach efforts expanded further through ACDA, which initiated programs in science museums such as the Bogota Planetarium and Maloka. These initiatives were replicated across the country in other science museums, and in 2009, Parque Explora in Medellin city, in collaboration with professionals from the Universidad de Antioquia (University of Antioquia), launched an outreach activity called the Journal Club. This initiative became a space for discussion around astrobiology-related concepts and led to the creation of the Multidisciplinary Group for Studies in Biology and

Astrobiology (AMEBA), which brought together individuals interested in the field (Rodríguez, 2018).

It was not until 2009 that UNASB took a distinctive direction compared to other groups and associations, by incorporating not only outreach, but also research and education as core areas of work—thus establishing itself as a national reference (Grupo de Ciencias Planetarias y Astrobiología, 2017). One of the milestones of that year was the publication of the book: The Ontogeny of Evolutionary Thought (La ontogenia del pensamiento evolutivo) by one of the professors then leading the student club in astrobiology and theoretical biology. The book explores the relationship between evolutionary epistemology and biological evolution, serving as an academic foundation for the study of one of astrobiology's central questions: how life evolves (Andrade, 2009). Furthermore, in 2011, the article Planet Earth as a Receptacle of Life: An Ordinary Planet or a Rarity in the Universe? (El planeta Tierra como un receptáculo de vida: ¿Un planeta corriente o una rareza en el universo?) was published, contributing to philosophical and scientific reflection within the field of astrobiology (Portilla, 2011).

Planetary geology began to take its first steps in Colombia in 2012, with the development of an undergraduate thesis comparing the largest volcanic events on Earth with super-eruptions on Jupiter's moon Io (Tovar & Sánchez-Aguilar, 2012), at least as far as current records show. That same year, the Titan Planetary Sciences Club was consolidated, with the aim of conducting research, outreach, and education in the field. In 2013, collaborative efforts began between the Vida sin Fronteras Astrobiology Club and Titan, ultimately leading to the establishment of the Planetary Sciences and Astrobiology Group (GCPA). This group is currently endorsed by the Universidad Nacional de Colombia and the Laguna Scientific Corporation (Corporación Científica Laguna), recognized by the Colombian Ministry of Science, Technology, and Innovation (Minciencias), and serves as the Colombian node of the Latin American Astrobiology Network.

In the same year, 2013, a research line in astrobiology was established within the research group on Stress Physiology and Biodiversity in Plants and Microorganisms at the Universidad Nacional de Colombia—one of the highest-ranked research groups according to the Colombian Ministry of Science, Technology, and Innovation (MinCiencias). Likewise, the Physics and Computational Astrophysics Group (FACOM) at the Universidad de Antioquia incorporated an astrobiology research line. The consolidation of astrobiology and planetary geology within research groups recognized by the Ministry marked a milestone for the country, as these areas of study were now formally acknowledged by the national research system.

7.3 Science Outreach and Public Communication of Science

Outreach has been one of the fundamental pillars in the development of astrobiology and planetary geology in Colombia, particularly considering that the introduction of these fields in the country began through outreach activities, due to the general lack of knowledge among both the public and even academic communities. One of the major challenges was the frequent confusion of these scientific disciplines with pseudoscience such as ufology. Therefore, in recognition of the origins of these sciences in the country and the persistent gaps in access to knowledge across various regions and socioeconomic sectors, efforts in the social appropriation of knowledge, scientific outreach, and public communication of science remain essential.

As previously mentioned, ACDA began offering outreach talks in a variety of venues, some of which have continued for over 28 years—such as the Saturday astronomy gatherings at the Bogotá Planetarium. These sessions regularly address topics related to Astrobiology and Planetary Sciences (Fig. 7.2) (Planetario de Bogotá, 2022). Additionally, ACDA's long-standing contributions include numerous lectures delivered at events such as the annual meeting of the Red de Astronomía de Colombia—RAC (Colombian Astronomy Network) and various astronomy and space science festivals held throughout the country.

One notable initiative within the framework of space science festivals is the Astrobiology Festival, developed by the Universidad de La Sabana (University La Sabana) and the Secretariat of Education of the Municipal Government of Cota. The first edition of the festival took place in 2020 and was conducted virtually, featuring activities such as workshops on extremophiles, impact craters, terrestrial analogs, and the origin of life, as well as lectures on a wide range of astrobiology topics. The Astrobiology Festival held its second edition in 2021 and a third in 2022 (Alcaldía Municipal de Cota, 2021). One of the most remarkable aspects of this festival was the participation of multiple institutions and, notably, the active involvement of students from local schools. These students were not only attendees, but also facilitators of activities and presenters of the various educational products developed in their academic seedbeds. This dynamic sparked interest and curiosity not only among the students themselves but also among the general public (Leal, 2024). Another festival organized by the University de La Sabana is the Astro-Sciences Festival, held in the municipality of Chía, Cundinamarca, which covered a wide range of topics (Fig. 7.3) (Universidad de La Sabana, 2023).

Among the strategies for the social appropriation of knowledge is the development of discussion spaces centered around science, where topics related to astrobiology and planetary sciences are addressed. One such space is Café y Ciencia (Coffee and Science), organized by the Planetary Sciences and Astrobiology Group (GCPA) since 2016. This initiative emerged as a means of disseminating astrobiology and planetary geology, adapting to the Colombian context the French strategy known as Café Scientifique. Under this format, every Tuesday from 6:00 p.m. to 8:00 p.m., a scientific topic is presented by an expert speaker, followed by an open discussion with the audience, all while enjoying coffee and cookies. With over 60 in-person sessions held since August 2016, and a subsequent transition to virtual meetings during the COVID-19 pandemic in 2020, the activity succeeded in gathering hundreds of attendees and dozens of speakers. It became a highly diverse and popular outreach platform to explore a wide range of scientific topics, enriched in each edition by current debates, valuable historical perspectives, and insights into

Fig. 7.2 Selected ACDA lecture series on astrobiology and planetary sciences, held as part of the Saturday astronomy gatherings. (Source: ACDA archive)

Fig. 7.3 Promotional posters for the festivals held in the Sabana Centro region, organized by the University of La Sabana. (Source: GCPA archive)

future developments across multiple fields. The activity resumed in 2023 with a new schedule, hence its current name: Café y Ciencia. These discussions aim to encourage participation from a wide variety of audiences and guests, recognizing in age, cultural, gender, and epistemic diversity an opportunity to co-construct knowledge (Grupo de Ciencias Planetarias y Astrobiología, 2016).

In addition to these initiatives, several strategies have been successfully developed in the digital realm. One such example is the digital capsule series Orígenes: Señales de la vida en el universo (Origins: Signs of Life in the Universe), produced by the Bogotá Planetarium. This series explores various aspects related to life and living systems in the cosmos. The videos incorporate interactive elements and diverse methods of content presentation, making astrobiology topics fully accessible to all audiences. Some of the subjects covered in the series include Life on Earth (Planetario de Bogotá, 2021a), Natural Satellites (Planetario de Bogotá, 2021b),

Titan and Enceladus (Planetario de Bogotá, 2021c), and Exoplanets (Planetario de Bogotá, 2021d).

Other outreach strategies have included a variety of public events featuring lectures and workshops aimed at the general public. Notable examples include activities organized by the Astrobiology Seedbed of the National University of Colombia, such as the Astrobiology Workshop held at the Luis Ángel Arango Public Library in 2009; the lecture Astrobiology: Life in the Cosmos from a Scientific Perspective in 2013 at the Bogotá Planetarium; talks on extreme environments and the search for extrasolar planets during the 17th Astronomy Festival of Villa de Leyva in 2014; as well as lectures titled Life in Extreme Environments and Prebiotic Chemistry and the Origin of Life, presented at the San Agustín Cloister in 2014; and the webinar Let's Talk About Astrobiology, held in 2020 in collaboration with the Peruvian Scientific Society of Astrobiology.

In addition to these public activities, institutions such as the Medellin Planetarium have also contributed through their Science on a Bicycle program, which has featured talks on planetary sciences and astrobiology. One such lecture was Living on Moons: Other Habitable Worlds, presented in 2021. Similarly, in 2020, the museum and astronomical observatory of the Metropolitan Technological Institute of Medellin hosted events on topics such as Extremophiles: A Mirror to Find Life on Other Planets and Fire and Ice Volcanoes in the Solar System. Along the same lines, in 2020, the University of La Sabana launched a webinar series on astrobiology, addressing themes such as Places in the Solar System to Search for Life and Extremophiles.

Over time, as numerous professionals in fields such as science, engineering, and health sciences had already participated in outreach-oriented events, new initiatives began to emerge in Colombia to promote the circulation of specialized knowledge. Some of these were led by the Planetary Sciences and Astrobiology Group (GCPA), including Astrogeology Day in 2014; the Planetary Sciences and Astrobiology Seedbed Meeting in 2019; and the Colombian Meeting on Women and Girls in Science, which began in 2019 and has since held seven editions. This event consistently features a dedicated space for discussing the role of women in astrobiology and planetary sciences, with its third edition entirely devoted to Mars. Another specialized event is Asteroid Day Colombia, first held in 2019, which has now completed seven editions. Additionally, the Workshop on Planetary Sciences and Astrobiology held its first edition in 2024 and its second in May 2025 (Corporación Científica Laguna, 2025a).

Other events have also emerged in various regions of the country, such as the International Congress of Astrobiology, which has held six editions and has been organized by institutions including Colegio Vermont (Universidad Nacional de Colombia, 2010), Parque Explora and the Medellin Planetarium, the University of Antioquia, the Colombian Institute of Astrobiology Corporation (Corporación Instituto de Astrobiología de Colombia) (Planetario de Medellín, 2012), the University of Manizales, the University of Atlántico, and the Cooperative University of Colombia. Another notable event was the First District Meeting on Astronomy and Astrobiology, organized by the Francisco José de Caldas District University in

Fig. 7.4 Covers of the five issues of Vida sin Fronteras published to date. (Source: GCPA archive)

Bogotá in 2023, which facilitated the engagement of biology, physics, and chemistry students with these topics (Universidad Distrital Francisco José de Caldas, 2023a). More recently, in April 2025, the Workshop on Astrochemistry and Astrobiology was held by the University of Valle (Universidad del Valle, 2025).

In the realm of public communication of science, the Planetary Sciences and Astrobiology Group (GCPA) has undertaken efforts to develop a strategy that ensures written dissemination. The objective has been to provide astrobiologists and planetary scientists with a platform to read and share their work in their native language. This led to the creation of the journal Vida sin Fronteras (Life Without Borders) in 2011, which has published five issues to date (Fig. 7.4) (Corporación Científica Laguna, 2025b). The journal is currently undergoing a redesign aimed at enabling publication in both Spanish and English, thereby ensuring broader accessibility of these academic contributions to the international scientific community. Another notable initiative in written science outreach was a series of columns published in the national newspaper El Tiempo under the title Wonders of the Solar System, which explored various aspects of planetary geology (Tovar, 2019). It is also important to highlight the publication of outreach books such as *Astrobiología: Un universo de vida* (Astrobiology: A Universe of Life) (Bueno & Moreno, 2011) and *La Incertidumbre de la Vida en el Cosmos* (The Uncertainty of Life in the Cosmos), published by the Asociación de Estudios Astronómicos ACDA. The latter was the result of decades of discussions held by the association's astrobiology group (Comisión Colombiana de Exobiología ACDA, 2018).

Another important avenue for astrobiology and planetary science outreach includes dedicated museum spaces, particularly interactive exhibition halls. One such space is the Astrobiology Hall at the Medellín Planetarium, inaugurated just over a decade ago (Planetario de Medellín, n.d.). Likewise, the interactive halls of the Bogotá Planetarium offer a narrative centered on life and living systems in the cosmos. These halls, renovated in 2022, together constitute the only museum in Latin America exclusively dedicated to exploring astrobiology and planetary sciences.

The first hall, titled We Are the Universe, addresses the scales of the cosmos and how humanity has interpreted its origins. The second, Understanding Life, explores multiple human approaches to comprehending the physical, geological, chemical, and biological processes that have made Earth a habitable planet. Finally, the third hall, The Future of Life, invites visitors to consider the possibility that life is an emergent property of matter that may have arisen in various places and times throughout the universe (Regalías Bogotá, 2024).

To activate the Bogotá Planetarium's interactive halls, the experience Living Universe was designed. Through full-dome projections and guided visits to exhibition devices—such as those explaining the Drake Equation, extremophiles, the origin of life, and geological time, among others—visitors are introduced to key concepts of astrobiology (Instituto Distrital de las Artes, 2022).

7.4 Education

From an educational perspective, various efforts have been made in Colombia to foster new scientific vocations in the field of planetary sciences and astrobiology. One of the earliest approaches has focused on engaging children, teenagers, and young students in basic education. Since 2015, the Bogotá Planetarium, in collaboration with the District Secretariat of Education, has implemented the Centers of Interest in Astronomy project. Beginning in 2021, astrobiology was adopted as a transversal and transdisciplinary theme within this initiative.

Astrobiology was chosen as a focal topic for early-age education because it enables the integration of concepts related to life and its implications, facilitates an understanding of our environment, and promotes reflection on how we interact with and study the world around us. Furthermore, these topics allow children to engage in meaningful discussions and connect with their own socio-environmental contexts (Secretaría de Educación del Distrito, 2023a).

To implement astrobiology as a central theme within the Centers of Interest in Astronomy, classroom discussions are framed within a school-based scientific inquiry process, developed according to the inquiry-based methodology and the pedagogical model of the Bogotá Planetarium (Leal et al., 2023a). In 2021, a total of 5323 students in Bogotá engaged with astrobiology-related topics (Secretaría de Educación del Distrito, 2021); in 2022, that number rose to 16,463 (Secretaría de educación del Distrito, 2021, 2022); in 2023, it reached 19,942 (Secretaría de educación del Distrito, 2022, 2023b); and in 2024, 6857 students participated in the program (Secretaría de Educación del Distrito, 2024). By 2025, the initiative aims to reach 12,000 students with content centered on astrobiology (Secretaría de Educación del Distrito, 2025).

Through these astrobiology-based processes, students develop essential scientific skills such as observation, inquiry, hypothesis formulation, critical thinking, and argumentation. As a result of the integration of astrobiology into the Centers of Interest in Astronomy, the Bogotá Planetarium has published four open-access educational booklets, which can be freely used by teachers and students across the Spanish-speaking world (Instituto Distrital de las Artes, 2021a, 2021b, 2021c, 2021d).

However, the Bogotá Planetarium's approach to astrobiology goes far beyond the aforementioned activities with basic education students. Since 2021, astrobiology has been incorporated as a transversal thematic axis across all of its pedagogical initiatives, a framework that will be maintained through at least 2026 and may be extended depending on the outcomes (Leal et al., 2023a, 2023b).

Within this context, the Planetarium began offering the Astrobiology for Enthusiasts course in 2019, in collaboration with the Planetary Sciences and Astrobiology Group (GCPA) (Grupo de Ciencias Planetarias y Astrobiología, 2019; Secretaría de Cultura, Recreación y Deporte, 2019). This course has since been offered annually in 2020, 2021, 2023, 2024, and 2025 (Planetario de Bogotá, 2020a, 2020b, 2021e, 2023a, 2024a, 2025).

In addition to this, the Planetarium has developed courses on various planetary science topics, including the Solar System and Exoplanets and Planetary Geology (Planetario de Bogotá, 2021f, 2023b, 2024b). Similarly, the Medellín Planetarium also offers astrobiology courses for both adults and children (Parque Explora, 2024).

In addition to the public courses previously mentioned, university-level courses have also emerged. One example is the elective course in Astrobiology offered by the Biology Program at the Francisco José de Caldas District University. This course is open to students across the university and covers key astrobiological concepts, including practical sessions designed to explore core topics in the field. As part of this academic track, an elective course in Geobiology has also been developed, offering a complementary and in-depth framework for students interested in astrobiology (Universidad Distrital Francisco José de Caldas, 2023b).

Moreover, the same university has offered short courses in planetary geology, such as the 2023 workshop Terrestrial Analogs and Their Relevance in Planetary Geology and Astrobiology, providing foundational training in this area.

It is important to highlight that this is not the only university offering courses in these fields of knowledge. In 2016, the National University of Colombia hosted the Alumni Lecture Series in Planetary Sciences and Astrobiology, where undergraduate students and alumni received foundational training in these disciplines. Additionally, in June 2019, as part of Research Month, the same university offered the course Research Methodology: Its Application in Planetary Sciences and Astrobiology in the city of Leticia.

Meanwhile, the Environmental Engineering Program at Universidad El Bosque offers the elective course Mars: New Challenges in Planetary Exploration. The course aims to promote the development of transversal skills and strengthen competencies such as critical thinking, creativity, and teamwork. It is structured around questions such as, "What is your proposal for adapting and developing your potential adaptability on Mars?"—placing university students from various academic disciplines in a scenario that simulates the possible arrival of humans on the Red Planet.

Additionally, the University of Antioquia has begun promoting a publicly accessible astrobiology course, and in 2024, the University of the Andes offered a course titled Astrobiology, open to the general public (Universidad de los Andes, 2024). This course is distinct from the elective course of the same name that has been part of the university's Physics Program since 2013. University of Valle has offered, since 2024, an elective course within its Chemistry Program titled Introduction to Astrochemistry. Similarly, in 2023, the University of Santiago de Cali offered a diploma course on Space Sciences and the Earth System, organized by its Faculty of Education.

As part of the academic training of university students, the development of student research seedbeds has become evident. The oldest of these is affiliated with the Planetary Sciences and Astrobiology Group (GCPA), whose origins date back to 2004—well before the official establishment of the research group itself. Over the years, students from diverse disciplines including biology, geology, physics, engineering, health sciences, and education have participated in this initiative.

More recently, other astrobiology, astrochemistry, and astronomy clubs have emerged, such as AstroLuca, a student research group in astrochemistry and astrobiology established in 2020 under the leadership of institutions including Universidad Libre (Cali campus), Universidad Santiago de Cali, Universidad del Valle, and the Cali School of Astronomy (EAC). This initiative was launched in response to the growing interest among faculty and students from fields such as basic sciences, education, health, and engineering, particularly given that, until then, these areas of knowledge had not been formally developed in the city of Cali. Additionally, the University of Antioquia, in collaboration with the Medellin Planetarium, continues to offer activities through the Multidisciplinary Group for Studies in Biology and Astrobiology (AMEBA).

7.5   Scientific Research

One of the most prominent research lines in astrobiology at the global level is the study of extreme environments, as these allow scientists to understand how life adapts to adverse conditions—insights that, in turn, help to develop models for the search for life elsewhere in the universe (Leal et al., 2015). Colombia has not been an exception in this research line, particularly considering that it is one of the most biodiverse countries in the world (Arbeláez-Cortés, 2013), as well as country rich in geodiversity (Gerstner et al., 2024). Colombia is home to a wide variety of environments, including thermal springs (Alfaro et al., 2021), tropical dry forests (Golley et al., 1969), snow-capped mountains (Ruiz et al., 2008), and glaciated volcanoes (Huggel et al., 2017), among others.

As a result of the aforementioned developments, several research projects have been carried out in Colombia. One such project is Characterization of Functional Groups of Thermophilic Microorganisms Present in Thermal Springs of the Machín-Cerro Bravo Volcanic Complex (Colombia) (Sánchez, 2024a, 2024b). This project, funded by the Administrative Department of Science, Technology, and Innovation (Colciencias), now the Ministry of Science, Technology, and Innovation (MinCiencias), aimed to characterize microorganisms that could serve as models for understanding the early stages of life, whether on primitive Earth or under ancient, warmer conditions on rocky planets such as Mars.

Another noteworthy research initiative is the project titled Characterization of Natural and Human-Altered Extreme Environments in Colombia with Potential for

Astrobiology, Planetary Sciences, and Microbial Ecology (Sánchez, 2019). This initiative has been linked to seedbed projects funded through the Project Management Program, such as Preliminary Identification of Extremophilic Microorganisms Associated with Xerophytic Environments in the Municipality of Villavieja, which explored xerophilic microorganisms in the Tatacoa Desert, Colombia (Bolivar et al., 2021), as well as studies involving cave environments (Buitrago et al., 2023).

However, Colombian research on extreme environments with implications for astrobiology and planetary sciences is not limited to locations within the country. Within the framework of international collaborations established through the Colombian Antarctic Program, scientific investigations are also being carried out in Antarctica—one of the most suitable regions on Earth for research in these fields. Research projects such as (1) Characterization of Functional Groups of Psychrophilic Microorganisms Present in Frozen Sediments of the Antarctic Peninsula, and (2) Antarctica as a Model for the Development of Astrobiology in the Nevado del Ruíz (Colombia) and the Return of Its Advances to Antarctic Science, researchers have been able to explore various types of microorganisms of astrobiological interest in both Antarctic environments and Colombian sites that share environmental similarities with Antarctica (Leal et al., 2025a; Acevedo-Barrios & Olivero-Verbel, 2021; Acevedo-Barrios et al., 2022a, 2022b, 2023, 2024a, 2024b, 2025; Bolaños et al., 2023).

Another line of research developed in Colombia focuses on the study of extreme environments from a planetary perspective, considering them as potential planetary analogs for rocky bodies such as Mars or the Moon. Notably, within the framework of previously mentioned projects, Deception Island has been evaluated as a potential multifunctional Martian analog. Similarly, the research project titled Assessment of the Ecological, Geological, Geochemical, and Climatological Characteristics of Gorgona Island and the Nevado del Ruíz Volcano as Terrestrial Analogs of Mars and Their Potential for Habitability and Analog Mission Development explores how these Colombian sites could be considered planetary analogs (Tovar et al., 2024) (Figs. 7.5 and 7.6). Such evaluations have also been carried out in Antarctic locations, including Deception Island (Leal et al., 2025). In the field of planetary geology, research has been conducted on rocky bodies such as Mercury (Ceferino, 2019), Europa (Lozada, 2018), Mars (Torres & Gutiérrez, 2024; Tovar et al., 2024), and Io (Tovar & Sánchez-Aguilar, 2012).

Another line of approach to astrobiological research stems from the field of chemistry, through studies that seek to understand environmental pollutants present in potentially analogous sites and their implications for planetary bodies such as Mars (Acevedo-Barrios et al., 2025). This line of inquiry also includes investigations into the relevance of astrochemistry for estimating the possibility of life in the universe (Pelegrín & Guerrero-Caicedo, 2023) and the perception of astrochemical knowledge in Colombia (García et al., 2023). These chemical relationships are directly linked to biology, either through the study of microorganisms tolerant to environmental pollutants (Acevedo-Barrios et al., 2024b) or through the

Fig. 7.5 Fieldwork at the Nevado del Ruíz Volcano. This site has been evaluated as one of the two terrestrial analogs studied to date in Colombia. Geomorphology and Geochemistry display similarities with Acidalia Planitia on Martian surface. (Credit: David Tovar, 2024)

Fig. 7.6 Fieldwork on Gorgona Island, located in the Pacific Ocean. This island is characterized by the presence of mafic and ultramafic rocks, with komatiites (above) being of particular interest in its evaluation as Colombia's second terrestrial analog site (Geochemical analog to Syrtis Major, Mars). (Credit: David Tovar, 2024)

understanding of biogeochemical processes associated with the origin of life (Reyes et al., 2024).

Another extensively studied area in the country relates to education in astrobiology and planetary sciences. Research in this field has been conducted at various school levels as well as in non-formal learning environments, demonstrating that these topics can support the integration and appropriation of scientific concepts while also inspiring children and young people to pursue science and develop scientific skills (Gil et al., 2023; Muñoz, 2020). Beyond these educational dimensions, other human-centered aspects have also been addressed, particularly with regard to the potential for human habitability on Mars. These include considerations of food systems, infrastructure design, and socio-political factors (Leal et al., 2021).

Additionally, one of the first milestones recorded in the field of Astrobiology research was the publication of one of the earliest research books in Spanish: *Temas selectos en astrobiología* (Selected Topics in Astrobiology). This book underwent a thorough peer-review process for each of the most relevant subjects in the field. It marked a significant departure from previously published works, as its target audience was scientists and scientists in training (Leal et al., 2015).

Finally, it is important to acknowledge that the studies presented here represent only a portion of the efforts currently underway in the country; other relevant work may have been conducted or may still be in progress.

7.6   Perspectives

As demonstrated throughout this chapter, the development of astrobiology and planetary sciences in Colombia has shown significant growth over the past decades, establishing itself as a multidisciplinary field that integrates research, education, public outreach, and international cooperation. Nevertheless, future projections call for the strengthening of various strategic areas to position the country as a regional leader in these disciplines.

One of the main priorities lies in reinforcing international collaborative networks, such as the Latin American Astrobiology Network, in which Colombia is represented by the Planetary Sciences and Astrobiology Group (GCPA). Another important step is the potential creation of advanced academic programs in these fields within the region, aimed at ensuring the training of highly qualified human capital capable of leading research initiatives and actively participating in global scientific missions.

It is also essential to strengthen specialized scientific infrastructure. Although the National University of Colombia hosts the Laboratory of Astrobiology, Planetary Sciences, and Microbial Ecology, and the University of Atlántico houses its own Astrobiology Laboratory, further investment is needed to support the appropriation of advanced technologies for planetary analog studies, the simulation of extreme environments, and the development of technologies for space exploration. Such facilities would not only enhance the country's research capabilities but also create opportunities for participation in international projects, including future exploration missions to Mars, the Moon, or other bodies within the solar system.

In the realm of education and the social appropriation of knowledge, it is crucial to sustain and expand initiatives that bring astrobiology and planetary sciences closer to diverse audiences, particularly in remote areas and communities that have traditionally been excluded from access to scientific knowledge. Integrating these topics into school and university curricula, alongside the use of digital technologies and innovative pedagogical strategies, will help to inspire new scientific vocations and foster a more informed and engaged citizenry committed to space exploration and development. Colombia also holds a comparative advantage due to its geodiversity, which provides unique settings for the study of planetary analogs and extreme environments. The systematic exploration of these environments, in synergy with public policies that recognize the strategic importance of space sciences, could position the country as a natural laboratory for astrobiology and planetary geology.

For this reason, the strengthening of science and technology policy oriented toward the space sector is an urgent necessity. The creation of specific national programs, the promotion of funding for interdisciplinary projects, and the establishment of a dedicated research agenda in planetary sciences and astrobiology will not only advance fundamental scientific knowledge but also enable the development of technologies with impact across sectors such as health, agriculture, environmental management, and industrial innovation. Supporting such decision-making processes will require continued research that demonstrates the need to develop this scientific ecosystem.

Declarations

Competing Interests The authors declare no competing interests.

References

ACDA. (2022). *Sobre ACDA*. Retrieved January 25, 2025, from https://www.acda.info/#portfolio. Consultado el 25 de enero de 2025 a las 18:23.

Acevedo, R. D., Rocca, M. C., & García, V. M. (2014). Colombia. In *Catalogue of meteorites from South America* (SpringerBriefs in earth system sciences). Springer. https://doi.org/10.1007/978-3-319-01925-3_6

Acevedo-Barrios, R., & Olivero-Verbel, J. (2021). Perchlorate contamination: Sources, effects, and technologies for remediation. *Reviews of Environmental Contamination and Toxicology, 256*, 103–120. https://doi.org/10.1007/398_2021_66

Acevedo-Barrios, R., Rubiano-Labrador, C., & Miranda-Castro, W. (2022a). Presence of perchlorate in marine sediments from Antarctica during 2017–2020. *Environmental Monitoring and Assessment, 194*(2), 102. https://doi.org/10.1007/s10661-022-09765-4

Acevedo-Barrios, R., Rubiano-Labrador, C., Navarro-Narvaez, D., Escobar-Galarza, J., González, D., Mira, S., Moreno, D., Contreras, A., & Miranda-Castro, W. (2022b). Perchlorate-reducing bacteria from Antarctic marine sediments. *Environmental Monitoring and Assessment, 194*, 654. https://doi.org/10.1007/s10661-022-10328-w

Acevedo-Barrios, R., Hernández Rocha, I., Puentes Martínez, D., Rubiano-Labrador, C., Pasqualino, J., Chavarro-Mesa, E., & De La Parra Querra, A. (2023). Psychrobacter sp: Perchlorate reducing bacteria, isolated from marine sediments from Margarita Bay, Antarctica. In *LACCEI international multi-conference for engineering, education, and technology*. https://doi.org/10.18687/LACCEI2023.1.1.995

Acevedo-Barrios, R., Rubiano-Labrador, C., Altamar Mercado, H., Villalba, J. L., Monroy-Licht, A., Carranza-López, L., Leal, M. A., & Tovar, D. (2024a). Identification of tardigrades from the Half Moon Island, Antarctic. In *LACCEI international multi-conference for engineering, education, and technology*. https://doi.org/10.18687/LACCEI2024.1.1.445

Acevedo-Barrios, R., Tirado-Ballestas, I., Bertel-Sevilla, A., Cervantes-Ceballos, L., Gallego, J. L., Leal, M. A., Tovar, D., & Olivero-Verbel, J. (2024b). Bioprospecting of extremophilic perchlorate-reducing bacteria: Report of promising Bacillus spp. isolated from sediments of the bay of Cartagena, Colombia. *Biodegradation, 35*(5), 601–620. https://doi.org/10.1007/s10532-024-10079-0

Acevedo-Barrios, R., Puentes Martínez, D. A., Hernández Rocha, I. O., Rubiano-Labrador, C., De la Parra-Guerra, A. C., Carranza-López, L., Monroy-Licht, A., Leal, M. A., & Tovar, D. (2025). Perchlorate in antarctica, origin, effects, treatments, environmental fate, and astrobiological perspectives: A review. *International journal of Environmental Science and Technology, 22*(5), 3855–3872. https://doi.org/10.1007/s13762-024-06004-w

Alcaldía Municipal de Cota. (2021). *Conocimiento*. Retrieved February 20, 2025, from https://www.cota-cundinamarca.gov.co/NuestraAlcaldia/SaladePrensa/Paginas/astrobiologia2021.aspx#gsc.tab=0

Alfaro, C., Rueda-Gutiérrez, J. B., Casallas, Y., Rodríguez, G., & Malo, J. (2021). Approach to the geothermal potential of Colombia. *Geothermics, 96*, 102169. https://doi.org/10.1016/j.geothermics.2021.102169

Andrade, E. (2009). *La ontogenia del pensamiento evolutivo*. Universidad Nacional de Colombia.

Arbeláez-Cortés, E. (2013). Knowledge of Colombian biodiversity: Published and indexed. *Biodiversity and Conservation, 22*, 2875–2906. https://doi.org/10.1007/s10531-013-0560-y

Bolaños, J., Buitrago, J., Leal, M., Tovar, D., Ruız, E., & Sánchez, J. (2023). Characterization of cultivable psychrophilic bacteria with phosphate-soluble activity and nitrogen fixation capacity, present in sediments of the Nevado del Ruiz (Caldas, Colombia). *Revista Mexicana de Astronomía y Astrofísica Serie de Conferencias (RMxAC), 55*, 65–67.

Bolivar, H., Méndez, Y., Sánchez, J., Leal, M. A., & Ruíz, E. (2021). Microorganismos xerófilos cultivables de la zona semiárida de la Tatacoa (Colombia). *Nova, 19*(36), 19–30.

Brungardt, M. P. (2000). Rebecca A. Earle, Spain and the Independence of Colombia, 1810–1825. *Colonial Latin American Historical Review, 9*(2), 301.

Bsdok, B., Altenberger, U., Concha-Perdomo, A. E., Wilke, F. D. H., & Gil-Rodríguez, J. G. (2020). The Santa Rosa de Viterbo meteorite, Colombia. New work on it's petrological, geochemical and economical characterization. *Journal of South American Earth Sciences, 104*, 102779.

Bueno, J., & Moreno, A. (2011). *Astrobiología: Un universo de vida*. Instituto de Astrobiología Colombia.

Buitrago, A. A., Sánchez, J., Leal, M. A., & Tovar, D. (2023). Microorganisms in Different Types of Caves in South America and Possible Survival Strategies. *Revista Mexicana de Astronomía y Astrofísica Serie de Conferencias (RMxAC), 55*, 125–125.

Ceferino, M. T. (2019). *Interpretación geológica de la Cuenca Caloris Planitia en Mercurio y las implicaciones de su proceso de formación en la zona antípoda*. Undergraduate geology thesis, Universidad Nacional de Colombia.

Comisión Colombiana de Exobiología ACDA. (2018). *La incertidumbre de la vida en el Cosmos*. Asociación Colombiana de estudios astronómicos ACDA.

Corporación Científica Laguna. (2025a). *Eventos Académicos*. Retrieved April 20, 2025, from en https://corpolaguna.org/eventos-academicos

Corporación Científica Laguna. (2025b). *Revista Vida Sin Fronteras*. Retrieved April 27, 2025, from https://corpolaguna.org/revista-vida-sin-fronteras

Escallón, M. E. (2015). "De lo que sin metáfora nos ha caído del cielo" Una presentación del Pequeño Museo del Aerolito de Santa Rosa de Viterbo. *Cuadernos de Música, Artes Visuales y Artes Escénicas, 10*(2), 217–224.

García, L., Guerrero-Caicedo, A., Pelegrín, J., Hernández, M., & Chaur, M. (2023). *Perceptions and Knowledge of Astrochemistry in the Colombian Chemical Community: Context analysis and experimental proposal*. Trabajo de investigación para optar al título de químico en la Universidad del Valle.

Gerstner, B. E., Blair, M. E., Bills, P., Cruz-Rodriguez, C. A., & Zarnetske, P. L. (2024). The influence of scale-dependent geodiversity on species distribution models in a biodiversity hotspot. *Philosophical Transactions of the Royal Society A, 382*(2269), 20230057. https://doi.org/10.1098/rsta.2023.0057

Gil, C., Leal, M. A., & Sánchez, J. (2023). *Diseño de una unidad didáctica de astrobiología para niños del ciclo I en los colegios del municipio de Ramiriquí (Boyacá-Colombia)*. Master's thesis, Universidad Nacional de Colombia.

Golley, F. B., McGinnis, J. T., Clements, R. G., Child, G. I., & Duever, M. J. (1969). The structure of tropical forests in Panama and Colombia. *Bioscience, 19*(8), 693–696. https://doi.org/10.2307/1294896

Grupo de Ciencias Planetarias y Astrobiología. (2016). *Martes de Café y Ciencia 2016-II*. Retrieved January 23, 2025, from https://www.youtube.com/watch?v=zqaUlr1XQ04&list=PLCCwVbPl kzCBTd9CiTQLpOSUGWqsve044&index=1

Grupo de Ciencias Planetarias y Astrobiología. (2017). *Informe a Facultad de Ciencias: Trayectoria Grupo de Ciencias Planetarias y Astrobiología GCPA*. Universidad Nacional de Colombia.

Grupo de Ciencias Planetarias y Astrobiología. (2019). *Curso de Astrobiología para aficionados*. Retrieved August 15, 2024, from https://www.youtube.com/watch?v=Rf5XwkgR66Q

Huggel, C., Ceballos, J. L., Pulgarín, B., Ramírez, J., & Thouret, J. C. (2017). Review and reassessment of hazards owing to volcano–glacier interactions in Colombia. *Annals of Glaciology, 45*, 128–136. https://doi.org/10.3189/172756407782282408

Instituto Distrital de las Artes. (2021a). *Cartilla Astrobiología Mi entorno terrícola Grados 1,2 y 3*. Retrieved September 15, 2024, from https://issuu.com/idartes/docs/cartilla_g123

Instituto Distrital de las Artes. (2021b). *Cartilla Astrobiología Mi entorno terrícola Grados 4 y 5*. Retrieved September 15, 2024, from https://issuu.com/idartes/docs/cartilla_g45

Instituto Distrital de las Artes. (2021c). *Cartilla Astrobiología Mi entorno terrícola Grados 6 y 7*. Retrieved September 15, 2024, from https://issuu.com/idartes/docs/cartilla_g67

Instituto Distrital de las Artes. (2021d). *Cartilla Astrobiología Mi entorno terrícola Grados 8 y 9*. Retrieved September 15, 2024, from https://issuu.com/idartes/docs/cartilla_g89

Instituto Distrital de las Artes. (2022). *Recorrido: Universo vivo, recorrido por el Planetario*. Retrieved April 15, 2025, from https://www.idartes.gov.co/es/agenda/recorrido/universo-vivo-recorrido-por-el-planetario-0

Leal, M. A. (2024). La Astrobiología: Una herramienta transformadora en la educación universitaria en Colombia. *Circular Astronómica Red de Astronomía de Colombia RAC, 995*, 8–9.

Leal, M. A., Escobar, J. F., Amaris, A., Saavedra, F., Tovar, D., Delgado, C., Reyes, R., Barriga, O., Doresty, F., Pinilla, M. A., Castañeda, Y., Pérez, C., Mantilla, F., Ojeda, O., & Álvarez, N. (2015). *Temas Selectos en Astrobiología*. Universidad Nacional de Colombia.

Leal, M. A., Tovar, D., Sánchez, J., Ceferino, M. T., Figueroa, L., Ríos, I., Leal, M., Reyes, R., Manjarrés, I., Quiñonez, E., Reyes, F., Orozco, M. C., Fung, Y., Barreto, C., Melgarejo, L. M, Matoma, S., Perico, D., Tovar, A., Antolinez, M., Arias, J., Martínez, E., Ballesteros, L., Franco, M., Rodríguez, A., Paez, S., Monsalve, S., Castaño, S., Parra, J., Castaño, A., Marín, A., Niño, A., García, G., Jiménez, J., Escobar, C., Ramírez, D., Bastidas, L., Nocua, I., Barriga, O., Barreto, N., Gahona, C., Rodríguez, J., Ruyardi, D., Porras, Y., Parada, A., Laverde, E., Ocampo, M., & Vega, V. (2021). *Mars Dichotomy: Prospects for human life on Mars*. https://doi.org/10.31223/X51S5J.

Leal, M. A., Molina, C., Valbuena, M., Guerra, Y., Carvajal, M., Tovar, D., Prada, N., Guevara, J., Ruíz, W., Guerrero, C., Sepulveda, K., Caicedo, D., Benavides, J. S., Ceferino, M. T., Montenegro, O., Altafulla, J. L., Cuartas, K., Pulido, C., Cano, F., & Cuervo, J. (2023a). *A multi-*

approach perspective for the attention of the diversity of the audiences in science museums: The pedagogical model of the Bogota Planetarium. Research Square Platform LLC. https://doi. org/10.21203/rs.3.rs-2967624/v1

Leal, M. A., Tovar, D., Valbuena, M., Guerra, Y., Sánchez, J., & Molina, C. (2023b). The transdisciplinary nature of astrobiology as a transversal axis of the educational processes at the Planetarium of Bogota. *Revista Mexicana de Astronom'ıa y Astrof'ısica Serie de Conferencias (RMxAC), 55,* 29–34.

Leal, M. A., Tovar, D., de Pablo, M. A., Bonilla, M. A., Leone, G., Tchegliakova, N., Sánchez, J., Molina, A., & San Martín, J. (2025). The potential of Deception Island, Antarctica, as a multifunctional Martian analogue of astrobiological interest. *International Journal of Astrobiology, 24,* e3. https://doi.org/10.1017/S1473550425000023

Leal, M. A., Tovar, D., Infante, A., Barriga, O., Ruíz, E., Sánchez, J., & Melgarejo, L. M. (2025a). Phosphate solubilization by microorganisms in pyroclastic material from Half Moon Island in Antarctica: Perspectives for astrobiology. *Polar Biology, 48*(1), 29. https://doi.org/10.1007/ s00300-025-03348-y

Lozada, M. (2018). *Mapa geomorfológico y distribución de terreno caótico conamara en satélite galileano Europa.* Undergraduate geology thesis, Universidad Industrial de Santander.

Morange, M. (2007). What history tells us X. Fifty years ago: The beginnings of exobiology. *Journal of Biosciences, 32*(5).

Muñoz, J. (2020). *El objeto virtual de aprendizaje (OVA) para la enseñanza de los microorganismos extremófilos con la astrobiología mediante la metodología ABP para estudiantes de secundaria.* Undergraduate education thesis. Universidad Pedagógica Nacional.

Paredes, L. S. (2016). *En las entrañas del país: Cien años de la Comisión Científica Nacional (1916–1940).* Museo Nacional de Colombia.

Parque Explora. (2024). *Vida en el Universo: Introducción a la astrobiología.* Retrieved January 20, 2025, from https://www.parqueexplora.org/tienda/cursos/vida-universo

Pelegrín, J. S., & Guerrero-Caicedo, A. (2023). Life Estimates in the Universe: Outstanding the Importance of Astrochemical and Astrobiological Parameters. *Revista Mexicana de Astronomía y Astrofísica Serie de Conferencias (RMxAC), 55,* 59–61.

Pita Pico, R. (2021). *El Reclutamiento de negros esclavos durante las Guerras de Independencia de Colombia 1810–1825.: El Reclutamiento de negros esclavos durante las Guerras de Independencia de Colombia 1810–1825* (Vol. 16). Academia Colombiana de Historia.

Planetario de Bogotá. (2020a).*Curso de Astrobiología para todos: Un acercamiento a un universo vivo.* Retrieved September 15, 2024, from https://planetariodebogota.gov.co/evento/ curso-astrobiolog-para-todos-un-acercamiento-un-universo-vivo

Planetario de Bogotá. (2020b). *Curso de Astrobiología: La vida y lo vivo.* Retrieved September 15, 2024, from https://planetariodebogota.gov.co/evento/curso-astrobiolog-vida-lo-vivo

Planetario de Bogotá. (2021a). *Señales de la vida en el universo: Vida en la Tierra.* Retrieved November 24, 2024, from https://www.youtube.com/watch?v=kJzO9qXBeT4

Planetario de Bogotá. (2021b). *Señales de la vida en el universo: Satélites naturales.* Retrieved November 24, 2024, from https://youtu.be/QELOq6IAVkg

Planetario de Bogotá. (2021c). *Señales de la vida en el universo: Titán y Encelado.* Retrieved November 24, 2024, from https://www.youtube.com/watch?v=mUaDMffIOtw

Planetario de Bogotá. (2021d). *Señales de la vida en el universo: Exoplanetas.* Retrieved November 24, 2024, from https://youtu.be/uvQcEk2bsMQ

Planetario de Bogotá. (2021e). *Curso de astrobiología.* Retrieved September 15, 2024, from https://www.planetariodebogota.gov.co/evento/curso-astrobiolog

Planetario de Bogotá. (2021f). *Curso Más allá, sistemas solares y exoplanetas.* Retrieved September 15, 2024, from https://planetariodebogota.gov.co/evento/curso-m-s-all- sistemas-solares-exoplanetas

Planetario de Bogotá. (2022). Retrieved February 10, 2025, from https://planetariodebogota.gov. co/evento/encuentros-sabatinos-astronom-acda.

Planetario de Bogotá. (2023a). *Curso—Astrobiología.* Retrieved September 15, 2024, from https:// www.planetariodebogota.gov.co/evento/curso-astrobiolog-0

Planetario de Bogotá. (2023b). *Curso—Geología Planetaria*. Retrieved September 15, 2024, from https://planetariodebogota.gov.co/programate/curso-geolog-planetaria

Planetario de Bogotá. (2024a). *Curso de Astrobiología*. Retrieved September 15, 2024, from https://www.planetariodebogota.gov.co/programate/curso-astrobiolog

Planetario de Bogotá. (2024b). *Curso Geología Planetaria*. Retrieved September 15, 2024, from https://planetariodebogota.gov.co/programate/curso-geolog-planetaria-0

Planetario de Bogotá. (2025). *Curso de Astrobiología*. Retrieved April 27, 2025, from https://planetariodebogota.gov.co/programate/curso-introducci-n-astrobiolog

Planetario de Medellín. (2012). *II Congreso Internacional de Astrobiología*. Retrieved October 31, 2024, from Disponible en https://www.flickr.com/photos/planetariomedellin/albums/72157631685512464/with/8051101136

Planetario de Medellín. (n.d.). *Astrobiología*. Retrieved April 21, 2025, from https://www.planetariomedellin.org/visitanos/astrobiologia

Portilla, J. G. (2011). El planeta Tierra como un receptáculo de vida: ¿Un planeta corriente o una rareza en el universo? *Acta Biológica Colombiana, 16*(3), 3–14.

Pulido Chaparro, S. C. (2015). De las reliquias patrimoniales a los recuerdos materializados en el Museo Nacional. *Memoria y Sociedad, 19*(38), 74–87.

Ramírez, J. E. (1949). The Meteorites of Santa Rosa de Viterbo, Boyacá, Colombia (730,059). *Contributions of the Meteoritical Society, 4*(15), 167–175. https://doi.org/10.1111/j.1945-5100.1949.tb00073.x

Ramírez, J. E. (1951). Sección Científica. *Theologica Xaveriana*, (1).

Regalías Bogotá. (2024). *Renovación Museográfica y de Espacios Planetario de Bogotá*. Retrieved April 20, 2025, from https://regaliasbogota.sdp.gov.co/es/proyectos/fcti/2021000100432/general

Reyes, R., Andrade, E., & Sánchez, J. (2024). *Evolución proto biótica de la ruta Wood-Ljungdahl (WL) en ambientes hidrotermales durante la vida temprana, eón Arqueano: Revisión, síntesis y modelo geoquímico*. Master's thesis, Universidad Nacional de Colombia.

Rodríguez, M. (2018). *Enseñanza y Aprendizaje de la Astronomía a través de instrumentos artesanales con estudiantes de grado décimo de la Institución Educativa Escuela Normal Superior de Neiva*. Undergraduate education thesis. Universidad Surcolombiana.

Rodríguez Prada, M. (2010). *La création du Musée national de Colombia (1823–1830): L'influence scientifique d'un modèle français*. Doctoral thesis in History of Art. University of Paris. Sorbonne.

Rodríguez Prada, M. (2014). Colecciones, registros e investigación: El caso del aerolito de Santa Rosa de Viterbo en el Museo Nacional de Colombia (1810–1830). In *Programa Ibermuseos* (pp. 97–110). Brasilia.

Ruiz, D., Moreno, H. A., Gutiérrez, M. E., & Zapata, P. A. (2008). Changing climate and endangered high mountain ecosystems in Colombia. *Science of the Total Environment, 398*(1–3), 122–132. https://doi.org/10.1016/j.scitotenv.2008.02.038

Sánchez, J. (2019). *Caracterización de Ambientes Extremos naturales e intervenidos de Colombia con potencial para la astrobiología, ciencias planetarias y ecología microbiana*. Retrieved March 20, 2025, from http://www.hermes.unal.edu.co/pages/Consultas/Proyecto.xhtml?idProyecto=45989

Sánchez, J. (2024a). *Caracterización de grupos funcionales de microorganismos Termófilos presentes en Fuentes Termales del complejo volcánico Machín-Cerro Bravo (Colombia)*. Retrieved March 20, 2025, from http://www.hermes.unal.edu.co/pages/Consultas/Proyecto.xhtml?idProyecto=30948

Sánchez, J. (2024b). *Volcan_cerro_machin. v1.91. No organization*. Retrieved March 20, 2025, from https://ipt.biodiversidad.co/permisos/resource?r=volcan_cerro_machin&v=1.91

Secretaría de cultura, recreación y deporte. (2019). *Curso de astrobiología para aficionados*. Retrieved September 15, 2024, from https://www2.culturarecreacionydeporte.gov.co/es/curso-de-astrobiologia-para-aficionados.

Secretaría de educación del Distrito. (2021). *Convenio Interadministrativo 2397115–2021 suscrito entre la Secretaría de Educación del Distrito y el Instituto Distrital de las Artes Idartes*.

Secretaría de educación del Distrito. (2022). *Convenio Interadministrativo 3808288–2022 suscrito entre la Secretaría de Educación del Distrito y el Instituto Distrital de las Artes Idartes.*

Secretaría de educación del Distrito. (2023a). *La astronomía llegó este año a 21 mil estudiantes de colegios oficiales de Bogotá.* Retrieved November 23, 2024, from https://www.educacionbogota.edu.co/portal_institucional/noticia/la-astronomia-llego-este-ano-21-mil-estudiantes-de-colegios-oficiales-de-bogota

Secretaría de educación del Distrito. (2023b). *Convenio Interadministrativo 4843561–2023 suscrito entre la Secretaría de Educación del Distrito y el Instituto Distrital de las Artes Idartes.*

Secretaría de educación del Distrito. (2024). *Convenio Interadministrativo 6083229–2024 suscrito entre la Secretaría de Educación del Distrito y el Instituto Distrital de las Artes Idartes.*

Secretaría de educación del Distrito. (2025). *Convenio Interadministrativo 7316101–2025 suscrito entre la Secretaría de Educación del Distrito y el Instituto Distrital de las Artes Idartes.*

Segura, M. (1995). *Itinerario del Museo Nacional de Colombia 1823–1994, t. I, Cronología* (vol. 2, pp. 31, 38–41, 205). Bogotá: Instituto Colombiano de Cultura—Colcultura, Museo Nacional de Colombia.

Tavera, M. A. (2015). *Evaluación e implementación de una propuesta de patrimonio geológico en el Parque Nacional Natural Los Nevados Cordillera Central colombiana.* Cuadernos Del Museo Geominero.

Tejfel, V. (2009). Gavriil Adrianovich Tikhov (1875–1960) a pioneer in astrobiology. *Proceedings of the International Astronomical Union, 5*(H15), 720–721.

Torres, M. C., & Gutiérrez, A. M. (2024). *Identificación de minerales hidratados y estimación de las edades de los cráteres de impacto presentes en Oxia Planum en Marte. Trabajo de grado para optar por el título de geólogas.* Universidad Industrial de Santander.

Tovar, D. (2019). *El Monte Olimpo/Maravillas del Sistema Solar.* Retrieved January 20, 2025, from https://www.eltiempo.com/vida/ciencia/el-monte-olimpo-columna-de-david-tovar-311964

Tovar, D., & Sánchez-Aguilar, J. J. (2012). Comparación entre los mayores eventos volcánicos de la Tierra y súper-erupciones en la luna Io de Júpiter. Geología Colombiana. *Edición X Semana Técnica de Geología e Ingeniería Geológica, 37*(1), 35–37.

Tovar, D., de Pablo, M. A., Leal, M. A., Tchegliakova, N., Molina, A., Leone, G., San Martín, J., Sánchez, J., Torres, A., & Bonilla, A. (2024). Gorgona Island as a geochemical terrestrial analog. *LPI Contributions, 3040*, 1539.

Universidad de La Sabana. (2023). *Festival de Astrociencias Chía.* Retrieved November 21, 2024, from https://www.unisabana.edu.co/noticias/al-dia/programese-para-un-festival-de-otro-planeta

Universidad de los Andes. (2024). *Astrobiología: Ciencia en la búsqueda de vida extraterrestre.* Retrieved January 13, 2025, from https://educacioncontinua.uniandes.edu.co/astrobiologia-ciencia-en-la-busqueda-de-vida-extraterrestre/p

Universidad del Valle. (2025). *Workshop en Astroquímica y Astrobiología.* Retrieved April 24, 2025, from https://workshopastro.correounivalle.edu.co/

Universidad Distrital Francisco José de Caldas. (2023a). *Semana de las Ciencias.* Retrieved September 20, 2024, from https://fnaturales.udistrital.edu.co/matematicas/sites/matematicas/files/archivo-publicaciones/2023-11/Agenda%20Semana%20de%20Las%20Ciencias%20 2023_0.pdf

Universidad Distrital Francisco José de Caldas. (2023b). *Exposición: 'Infografías de Geobiología' en el Planetario de Bogotá.* Retrieved September 15, 2024, from https://laud.udistrital.edu.co/universidad-distrital/exposicion-infografias-de-geobiologia-en-el-planetario-de-bogota

Universidad Nacional de Colombia. (2010). *Primer Congreso Internacional de Astrobiología.* Retrieved October 31, 2024, from https://agenciadenoticias.unal.edu.co/detalle/primer-congreso-internacional-de-astrobiologia

Universidad Nacional de Colombia. (n.d.). *Cuando la UNAL estudió el meteorito más famoso del país.* Retrieved March 13, 2025, from https://gestiondocumental.unal.edu.co/cuando-la-unal-estudio-el-meteorito-mas-famoso-del-pais

Ward, H. A. (1907). ART. I.—Colombian Meteorite Localities: Santa Rosa, Rasgata, Tocavita. *American Journal of Science (1880–1910), 23*(133), 1.

Chapter 8
Astrobiology in Cuba

Rolando Cardenas

Abstract This chapter offers a historical and analytical overview of the development of astrobiology in Cuba, with a focus on research, teaching, and public outreach initiatives led primarily by the Planetary Science Laboratory at Universidad Central "Marta Abreu" de Las Villas. Since the early 2000s, Cuban researchers have contributed to astrobiology through interdisciplinary studies on prebiotic chemistry, planetary habitability, solar-terrestrial interactions, and the effects of astrophysical radiation on ecosystems. The transformation of the Gravitation and Cosmology Group into the Planetary Science Laboratory marked a turning point in establishing astrobiology as a recognized scientific field in the country. Collaborative efforts with national and international partners, including the organization of the BioGeoSciences Conferences and contributions to Springer Proceedings, have positioned Cuba as an emerging contributor to global astrobiological research. The integration of environmental sciences and astrobiology, especially under resource-constrained conditions, is highlighted as a model for scientific advancement and societal relevance in developing countries.

8.1 Introduction

Astrobiology in Cuba has developed progressively over the past two decades, with its main institutional representation centered at the *Universidad Central "Marta Abreu" de Las Villas* (UCLV), located near the city of Santa Clara in central Cuba. The emergence of this field at UCLV has been closely associated with interdisciplinary research initiatives in cosmology, planetary sciences, and environmental studies, which have provided a fertile ground for the integration of astrobiological approaches. The transformation of traditional physics groups into more

R. Cardenas (✉)
Planetary Science Laboratory, Universidad Central "Marta Abreu" de Las Villas,
Santa Clara, Cuba
e-mail: rcardenas@uclv.edu.cu

D. Tovar, M. A. Leal (eds.), *Astrobiology and Planetary Sciences in Latin America*, https://doi.org/10.1007/978-3-032-01450-4_8

transdisciplinary teams reflects a national commitment to exploring fundamental scientific questions about the origin and distribution of life in the Universe, even within resource-limited academic environments.

Parallel to these efforts, researchers from the Institute of Geophysics and Astronomy in Havana have contributed to astrobiology-related themes through the study of Sun–Earth interactions. Although originally framed within the scope of heliophysics and atmospheric sciences, these studies have explored correlations between solar variability and biological effects on Earth, particularly in human health and ecological systems. Such research has served as a foundation for broader discussions on planetary habitability and the potential influence of stellar activity on life, both on Earth and in extraterrestrial contexts.

Together, these contributions reflect the growing presence of astrobiology in the Cuban scientific landscape. Through collaborations with international research networks, scientific conferences, and the training of undergraduate and graduate students, astrobiology in Cuba has evolved into a field with significant academic projection. The integration of fundamental and applied sciences has allowed researchers to explore questions about life in the universe while also engaging in knowledge production that is relevant across disciplines.

8.2 Research

In the decade 2001–2010 Dr. Rolando Cardenas, from the Gravitation and Cosmology Group of UCLV, introduced Astrobiology as a research direction in this group. In 2003 he started collaboration with Dr. Jorge Horvath, from the Astronomy Department of the Institute of Astronomy, Geophysics and Atmospheric Sciences of University of Sao Paulo (Brazil). They shared a (Cuban) Ph.D. student and several interesting publications were achieved (Martín et al., 2009a, 2009b, 2010, 2012).

On another hand, the relation Sun-Earth was treated by some researchers in a perspective which some people call Heliobiology. For instance, a bit prior to the beginning of this decade researchers from the above mentioned Institute of Geophysics and Astronomy looked at the relationship solar variability—asthmatic crises in persons (Del Pozo et al., 2000). Inspired by this, Dr. Gladys Casas and Dr. Rolando Cardenas advised the M. Sc. thesis in Applied Mathematics "Correlation between Solar and Geomagnetic Activities and Cardiovascular Diseases", presented in 2008 by Ibis Ramos.

In the decade 2011–2020, due to the high interdisciplinarity given by Astrobiology, the Gravitation and Cosmology Group was transformed into the Planetary Science Laboratory, where research in Theoretical Cosmology, Astrobiology and Environmental Sciences takes place up today. In that decade, some members of this laboratory became aware of the astrobiological importance of Loma del Capiro (in the outskirts of Santa Clara), as it is one of the places in which geologists have documented in Cuba fingerprints of the Chicxulub asteroid impact, the one which exterminated so many living species, including the dinosaurs. Thus, Dr. Osmel Martin from the Planetary Science Laboratory, advised a Ph.D. student (Noel Pérez) to research on the influence of this impact on both photosynthetic and chemoautotrophic primary producers (Pérez et al., 2013b, 2014, 2019, 2020). During this doctorate, life under other extreme environments was also studied (Pérez et al., 2013a). This Ph.D. student was co-advised by Jesús Martínez-Frias, President of the Spanish Network of Planetology and Astrobiology. In this decade, the Planetary Science Laboratory organized the 1st, 2nd and 3rd International Conferences on BioGeoSciences (years 2013, 2017 and 2019, respectively). Proceedings of the 2nd and 3rd Conferences were published by the prestigious scientific editorial Springer Nature (Cardenas et al., 2019, 2022). Of particular relevance for Astrobiology were the opening talks (Cardenas et al., 2019, 2022).

In the current decade, Dr. Rolando Cardenas continues doing research in Environmental Sciences (Oceanography, Limnology and Biophysics) and in Astrobiology-related topics, especially devising habitability metrics both for photosynthetic and chemoautotrophic species (Rodríguez-López et al., 2021a; Cardenas & Rodríguez-López, 2023). These metrics can in principle be applied in Astrobiology and in Environmental Sciences in general. He has consolidated his collaboration with Lien Rodríguez-López, his former Ph.D. student from the Centre for Environmental Research (EULA) at the University of Concepción (Chile). An important part of this collaboration is research on the influence of astrophysical sources of ionizing radiation on life on Earth (Rodríguez-López et al., 2021b, 2023; López-Águila et al., 2023). On another hand, Dr. Osmel Martín continues in the Planetary Science Laboratory and explores other topics in Astrobiology, such as the potential role of homochirality in the origin of life (Martín et al., 2023a, 2023b) and other aspects of prebiotic chemistry (Martín et al., 2021; Martin et al., 2022). The 4th International Conference on BioGeoSciences was organized (on November/2023), where there was plenty of interaction between astrobiologists and environmentalists. It is expected to publish its Proceedings again by Springer Nature, and as in the former version, in the series Springer Proceedings on Earth and Environmental Sciences.

8.3 Teaching

Since the early efforts to introduce astrobiology in Cuba, researchers from the Planetary Science Laboratory at Universidad Central "Marta Abreu" de Las Villas have actively supported undergraduate education in this field. In particular, Dr. Rolando Cárdenas, Dr. Osmel Martín, and Dr. Noel Pérez have incorporated astrobiology-related topics into elective courses for students in Physics, Chemistry, and Biology. These courses have offered students the opportunity to explore astrobiology from an interdisciplinary perspective, connecting astronomical, biological, and geophysical processes within a unified scientific framework.

This academic support has extended beyond the classroom. Several students have been guided in the development of their Bachelor of Science theses on astrobiology-related subjects. These projects have explored a variety of themes, including the habitability of exoplanets, the impact of solar activity on living systems, and the chemical conditions for the emergence of life. The supervision process has emphasized scientific rigor and critical thinking, allowing students to engage with current research questions and contribute to the growing academic conversation in the field.

Some of these undergraduate theses have resulted in publications in nationally recognized scientific journals (Gonzalez et al., 2013; López-Águila et al., 2013), reflecting both the quality of the research and the institutional support provided by the Planetary Science Laboratory. These contributions have helped position astrobiology as a relevant and inspiring area of study within undergraduate education in Cuba, fostering scientific curiosity and encouraging interdisciplinary dialogue among students and faculty alike.

8.4 Outreach

Around the year 2010, members of the Planetary Science Laboratory began participating actively in activities organized by amateur astronomer groups throughout Cuba—an engagement that continues to this day. These interactions have played a fundamental role in bridging the gap between academic research and public interest in the cosmos. Scientists from the Laboratory have joined observational gatherings, lectures, and informal science cafés, offering insights into contemporary topics such as planetary science, the conditions for life in the Universe, and recent developments in astrobiology. This ongoing collaboration has fostered a shared space for dialogue and learning, where both professionals and enthusiasts contribute to a vibrant and inclusive scientific culture.

Throughout the country, various amateur astronomy groups meet periodically to observe celestial phenomena, exchange knowledge, and discuss both classical and emerging topics in astronomy. Among these topics, astrobiology has gained growing attention, thanks in part to the interdisciplinary nature of the field and its philosophical implications about humanity's place in the Universe. The Wilhelm Herschel group in Caibarién and the Meridiano 80 group in Santa Clara have become particularly prominent within this community. Both groups have developed close ties with researchers from the Planetary Science Laboratory and have hosted events that highlight the relevance of astrobiology for both scientific understanding and public imagination. Activities have included skywatching sessions, public talks, and collaborative workshops that explore the intersection between space science and life sciences.

In parallel with these grassroots initiatives, national mass media has also contributed to the diffusion of astronomical and astrobiological knowledge in Cuba. Since 1997, the television program Pasaje a lo Desconocido (Ticket to the Unknown) has played a leading role in scientific outreach. Hosted on Cuban national television, the program has addressed a wide range of scientific topics, with astronomy occupying a central place in its content. Many of the most popular episodes have been devoted to issues related to astrobiology, such as the search for extraterrestrial life, habitable exoplanets, and cosmic evolution. The insightful commentary of astronomer Oscar Álvarez, from the Cuban Academy of Sciences, added depth and clarity to these discussions. The program remains on air to this day, serving as an influential platform for public science communication and as an important ally in promoting interest in astrobiology across Cuban society.

8.5 Final Considerations

At this specific point, it is worth highlighting the symbiosis between astrobiology (at least, in its purest conception) and the more environmental work developed by our group in recent years. Our experience shows that such integration can be an opportunity, especially for poor countries with low budgets where astrobiology could be considered a rather elitist specialty with little practical application. On the other hand, for teaching, this integration makes it possible to develop a more harmonious conception of students of basic and applied sciences that allows them not only to expand their scientific horizons, but also to find effective and socially useful ways to address many of the problems of the contemporary world.

Acknowledgments The author thank colleague Osmel Martín for useful comments and suggestions.

Declarations

Competing Interests The authors declare no competing interests.

References

Cardenas, R., & Rodríguez-López, L. (2023). Quantification of habitability in rocky planetary bodies. *Revista Mexicana de Astronomía y Astrofísica Serie de Conferencias (RMxAC), 55,* 1–9. https://doi.org/10.22201/ia.14052059p.2023.55.01

Cardenas, R., Nodarse-Zulueta, R., Perez, N., Avila-Alonso, D., & Martin, O. (2019). On the quantification of habitability: Current approaches. In R. Cardenas, V. Mochalov, O. Parra, & O. Martin (Eds.), *Proceedings of the 2nd international conference on BioGeoSciences. BG 2017.* Springer. https://doi.org/10.1007/978-3-030-04233-2_1

Cardenas, R., Perez, N., Martin, O., & Horvath, J. (2022). Opening talk: What pervades the milky way: The snotty or the stringy? Photoautotrophy or chemoautotrophy? In R. Cardenas, V. Mochalov, O. Parra, & O. Martin (Eds.), *Proceedings of the 3rd international conference on BioGeoSciences* (Springer proceedings in earth and environmental sciences). Springer. https://doi.org/10.1007/978-3-030-88919-7_1

Del Pozo, E., Gil, G., Valiente, J., Batule, M., Toledo, H., & Triolet, A. (2000). Induced effects of solar variability on asthmatic crises in adults. *Geofísica Internacional, 39*(4), 367–372.

Gonzalez, A., Cardenas, R., & Hearnshaw, J. (2013). Possibilities of life around alpha Centauri B. *Revista Cubana de Física, 30,* 81–83.

López-Águila, M., Cárdenas-Ortiz, R., & Rodríguez-López, L. (2013). On the habitability of exoplanets orbiting Proxima Centauri. *Revista Cubana de Física, 30*(2), 77–80.

López-Águila, M., Rodríguez-López, L., González-Rodríguez, L., Cardenas, R., & Borges-Márquez, J. (2023). Short-term effects of solar storms in phytoplankton photosynthesis. *Astrophysics and Space Science, 368*, 1–9. https://doi.org/10.1007/s10509-023-04162-w

Martín, O., Peñate, L., Alvaré, A., Cardenas, R., Horvath, J., & Galante, D. (2009a). Some possible constraints for life's origin. *Origins of Life and Evolution of Biospheres, 39*(6), 533–544.

Martín, O., Galante, D., Cardenas, R., & Horvath, J. (2009b). Short-term effects of gamma rays bursts on earth. *Astrophysics and Space Science, 321*, 161–167.

Martín, O., Cardenas, R., Guimarais, M., Peñate, L., Horvath, J., & Galante, D. (2010). Effects of gamma rays bursts in earth's biosphere. *Astrophysics and Space Science, 326*, 61–67.

Martín, O., Peñate, L., Cardenas, R., & Horvath, J. (2012). The photobiological regime in the very early earth and the emergence of life. In J. Seckbach (Ed.), *Genesis—In the beginning. Cellular origin, life in extreme habitats and astrobiology* (Vol. 22). Springer. https://doi.org/10.1007/978-94-007-2941-4_8

Martín, O., Leyva, Y., Suárez-Lezcano, J., Pérez-Castillo, Y., & Marrero-Ponce, Y. (2021). The minimal and the optimal size for two different types of encapsulated replicator systems. *Chinese Journal of Physics, 71*, 397–402. https://doi.org/10.1016/j.cjph.2021.03.012

Martin, O., Suarez-Lezcano, J., & Leyva, Y. (2022). Darwinian evolution from a generational point of view. In R. Cardenas, V. Mochalov, O. Parra, & O. Martin (Eds.), *Proceedings of the 3rd international conference on BioGeoSciences* (Springer proceedings in earth and environmental sciences). Springer. https://doi.org/10.1007/978-3-030-88919-7_14

Martín, O., Leyva, Y., Suárez-Lezcano, J., Pérez-Castillo, Y., & Marrero-Ponce, Y. (2023a). From a coenzyme-like mechanism to homochirality. *Biosystems, 227–228*, 104904. https://doi.org/10.1016/j.biosystems.2023.104904

Martín, O., Leyva, Y., Suárez-Lezcano, J., Pérez-Castillo, Y., & Marrero-Ponce, Y. (2023b). Inducing homochirality through intermediary catalytic species: A stochastic approach. *Astrobiology, 23*, 1083–1089. https://doi.org/10.1089/ast.2023.0004

Pérez, N., Cardenas, R., Martín, O., & Leiva, M. (2013a). The potential for photosynthesis in hydrothermal vents: A new avenue for life in the universe? *Astrophysics and Space Science, 346*, 327–331.

Pérez, N., Cardenas, R., Martín, O., & Rojas, R. (2013b). Modeling the onset of photosynthesis after the Chicxulub asteroid impact. *Astrophysics and Space Science., 343*, 7–10.

Pérez, N., Martín, O., & Cardenas, R. (2014). Evolución del proceso de fotosíntesis después del impacto del asteroide de Chicxulub. *Revista Cubana de Física, 31*(2E), E55–E57.

Pérez, N., Velazco-Vargas, J., Martín, O., Cardenas, R., & Martínez-Frías, J. (2019). The mass impacts on chemosynthetic primary producers: Potential implications on anammox communities and their consequences. *International Journal of Astrobiology, 18*(5), 440–444. https://doi.org/10.1017/S1473550418000411

Pérez, N., Martin, O., & Cárdenas, R. (2020). Potential changes on anammox activity after Chicxulub asteroid impact. In R. Cárdenas, V. Mochalov, O. Parra, & O. Martin (Eds.), *Proceedings of the 2nd international conference on BioGeoSciences. BG 2017*. Springer. https://doi.org/10.1007/978-3-030-04233-2_7

Rodríguez-López, L., González-Rodríguez, L., Cardenas, R., & Peñate, L. (2021a). Inclusion of ionizing radiation in a mathematical model for photosynthesis. *Radiation and Environmental Biophysics, 60*, 431–435. https://doi.org/10.1007/s00411-021-00918-6

Rodríguez-López, L., Cardenas, R., González-Rodríguez, L., Guimarais, M., & Horvath, J. (2021b). Influence of a galactic gamma ray burst on ocean plankton. *Astronomische Nachrichten, 342*, 45–48. https://doi.org/10.1002/asna.202113878

Rodríguez-López, L., González-Rodríguez, L., Díaz-García, T., & Cardenas, R. (2023). Influence of cosmic rays on environmental ecosystems. *Astronomische Nachrichten, 344*, e220125. https://doi.org/10.1002/asna.20220125

Chapter 9
Sociedad Mexicana de Astrobiología

Guadalupe Cordero-Tercero, Irma Lozada-Chávez, Hugo Beraldi-Campesi, Eduardo Piña, Sandra Ramírez Jiménez, Antígona Segura, and Roberto Vázquez

Abstract The Mexican Society of Astrobiology (Sociedad Mexicana de Astrobiología, SOMA) started in 2000 with the name of Sociedad Mexicana de Ciencias de la Vida en el Espacio (Mexican Society of the Sciences of Life in Space), and changed its name to Sociedad Mexicana de Astrobiología in 2002. The Society has been the core driver of the development of astrobiology in Mexico, promoting this discipline by organizing talks, national meetings and conferences and schools for Mexican and international researchers and students, as well as outreach activities. This chapter presents the history and activities of SOMA since its foundation.

G. Cordero-Tercero (✉)
Sección de Riesgos Espaciales, Instituto de Geofísica, Universidad Nacional Autónoma de México, Mexico City, Mexico
e-mail: gcordero@igeofisica.unam.mx

I. Lozada-Chávez
Faculty of Mathematics and Computer Science, Institute of Computer Science, Leipzig University, Leipzig, Germany

H. Beraldi-Campesi
Sociedad Mexicana de Astrobiología, A.C., Mexico City, Mexico

E. Piña
Dirección General de Divulgación de la Ciencia, Universidad Nacional Autónoma de México, Mexico City, Mexico

S. Ramírez Jiménez
Centro de Investigaciones Químicas, Universidad Autónoma del Estado de Morelos, Cuernavaca, Mexico

A. Segura
Instituto de Ciencias Nucleares, Universidad Nacional Autónoma de México, Mexico City, Mexico

R. Vázquez
Laboratorio de Astrobiología, Instituto de Astronomía, Universidad Nacional Autónoma de México, Ensenada, B.C., Mexico

D. Tovar, M. A. Leal (eds.), *Astrobiology and Planetary Sciences in Latin America*, https://doi.org/10.1007/978-3-032-01450-4_9

9.1 The Early History of Astrobiology in Mexico and the Origins of SOMA

The first studies of astrobiology in Mexico were led by Rafael Navarro-González (1959–2021) at the Instituto de Ciencias Nucleares (Universidad Nacional Autónoma de México, UNAM), where he founded the Laboratorio de Química de Plasmas y Estudios Planetarios (Laboratory of Plasma Chemistry and Planetary Studies) in 1996. Initially, his research was focused on the role of lightning in nitrogen fixation on the early Earth and Mars (Navarro-González et al., 1998), and the atmospheric chemistry of Saturn's moon Titan. Later, he studied the potential for prebiotic chemistry in the hypothetical hydrothermal vents in Jupiter's moon Europa, and protocols for detecting life on Mars. After conducting experiments with desert soils, similar to the Viking mission experiments, his research group found that this mission may have missed the detection of organic matter due to a flaw in the experimental design, which did not account for the presence of organic matter oxidants, such as perchlorates (Navarro-González et al., 2010). This discovery aided in refining NASA's plans for the Curiosity rover mission and Navarro-González was invited to collaborate with the SAM (Sample Analysis at Mars) instrument, aiming to identify organic compounds. Over the years, several students have worked in his laboratory and eventually graduated as chemists, biologists, and Earth scientists specialized in astrobiology. The first students formed in Navarro-González's laboratory were Sandra Ramírez, Antígona Segura, José de la Rosa, Lilia Montoya and Paola Molina, who are now active researchers and teachers working on astrobiology and, either past or present, members of SOMA.

In 2000, the physician Ramiro Iglesias Leal, the astronomer Miguel Ángel Herrera Andrade and the engineer José de la Herrán decided to promote space science and astronomy studies related to the possibility of finding life in the universe by founding a civil association called Sociedad Mexicana de Ciencias de la Vida en el Espacio. Rafael Navarro-González and Antígona Segura were later invited to participate in the Society, who then suggested that 'astrobiology' was a better description for the Society than 'sciences of life in space'. In 2001, the Society organized the I Reunión de la Sociedad Mexicana de Astrobiología (First Mexican Meeting of Astrobiology) at the Instituto de Ciencias Nucleares (UNAM); This activity was very well received by the students and helped to identify other researchers interested in astrobiology. The Society used its current name at this meeting and it was legally changed in 2002.

The purpose of the Society as described in its statutes[1] are:

1. To promote the study and development of sciences related to Astrobiology in Mexico.
2. To gather scientists, professionals and researchers who are interested in the study and development of Astrobiology to increase knowledge in this field.

---

[1] Sociedad Mexicana de Astrobiología (SOMA): https://www.soma.org.mx

3. To establish relationships with national and international organizations dedicated to Astrobiology and related sciences.
4. To disseminate, for educational purposes, in the student community and the general public of Mexico and abroad, the advances in the knowledge of Astrobiology.
5. To contribute to the training of human resources to the study and development of Astrobiology.
6. To take advantage of the studies and experiences derived from science in space, in the preservation of the terrestrial environment, its expressions of life and its natural ecological harmony.

The presidents of SOMA have been: Ramiro Iglesias Leal (2000–2003), Rafael Navarro González (2004–2010), Elva Escobar Briones (2010–2011), Antígona Segura Peralta (2011–2013, 2020–2021), Sandra Ignacia Ramírez Jiménez (2014–2015, 2024–2025), Roberto Vázquez Meza (2016–2017), Guadalupe Cordero Tercero (2018–2019) and Patricia Guadalupe Núñez Pérez (2022–2023).

9.2 SOMA and Current Research on Astrobiology in Mexico

The Society has around 50 members on average per year, most of them belonging to educational institutions whose activities include research, teaching and/or science communication in subjects such as astronomy, biology, chemistry, geology, medicine, mathematics, education, philosophy and history of science. Currently, some of the research fields related to astrobiology in Mexico are:

1. *Exoplanets*: Exoplanet detection and characterization techniques.
2. *Astrochemistry*: Formation and characterization of organic molecules in the interstellar medium (regions of star formation, planetary nebulae, etc.), and in carbonaceous chondrites.
3. *Extreme environments and extremophiles*: Studies on extremophile organisms and their adaptation to natural and controlled microenvironments, as well as the characterization of natural environments with these conditions.
4. *Origin of life*: Chemical and molecular evolution, first metabolisms and replicating systems, emergence of the first living beings and other biological entities (*e.g.*, viruses).
5. *Early evolution of life*: Characterization of the Last Universal Common Ancestor, evolution of early metabolisms and emergence of major biological transitions, such as syntrophy, multicellularity, symbiosis, eukaryogenesis and human intelligence.
6. *Biosignatures*: Numerical simulation of planetary atmospheres to determine the compounds produced by life that would be detectable in a planetary spectrum.
7. *Habitability and detection of life in solar system bodies*.
8. *Planetary habitability*: Effect of stellar activity (*e.g.*, flares, coronal mass ejections) on a planet's potential for the origin and evolution of life.

9. *Galactic habitable zone*: Temporal and spatial evolution of the metallicity of the galaxy and its relevance for the formation of rocky planets.
10. *Formation of planetary systems*: Formation and characterization of circumstellar disks.

More details about the astrobiology research developed in Mexico can be found in Cervantes de la Cruz et al. (2020). SOMA also promotes and makes publicly available a record of its science communication and research activities on diverse official social media and its website (Fig. 9.1). From February 2021 to February 2024, for instance, around 152.8 views on average have been recorded for the >70 video talks published on SOMA's YouTube channel (with over 600 subscribers and 36,000 views, Fig. 9.1a). Among the most watched talks (with 300–500 views) are: "¿Que podemos decir acerca del futuro de un mundo habitable?" (What can we say about the future of an habitable world?) by Dr. Eva G. Villaver Sobrino (Centro de Astrobiología, Spain); "Las arqueas y la vida en ambientes extremos" (Archaea and life on extreme environments) by Dr. Michel G. Santiago-Martínez (Pennsylvania State University, US); "La misión Perseverance y la búsqueda de vida en Marte" (The Perseverance mission and search for Life on Mars) by Dr. José A. Rodríguez Manfredi (Centro de Astrobiología, Spain); and "Homenaje al Dr. Rafael Navarro-González" (Tribute to Dr. Rafael Navarro-González) held at the III Congreso Latinoamericano de Astrobiología (III Latin American Astrobiology Conference).

Fig. 9.1 Official logo, social media (**a**) and website (**b**) of the Sociedad Mexicana de Astrobiología

9.3   Activities on Astrobiology in Mexico

9.3.1   *Academic Activities*

9.3.1.1   Meetings and Conferences

SOMA organized its first meeting at Instituto de Ciencias Nucleares, UNAM, in 2001 with the participation of a considerable number of students, professors, and researchers. Other meetings were organized at other universities (Fig. 9.2) until the tenth meeting in 2017. This meeting was organized with the Workshop on Planetary Astrophysics at the Universidad Autónoma de Nuevo León (UANL) in Monterrey, Mexico, and was the first to be recognized as a national conference of astrobiology. The most recent conference was hosted by the Instituto de Astronomía, UNAM, and the Universidad Autónoma de Baja California (UABC) in Ensenada, Baja California, in September 2023. Figure 9.2 shows the dates and places where SOMA has organized its meetings and conferences.

In each of these events, invited speakers from Mexico and abroad have had the opportunity to share their expertise with Mexican and few international students through short talks, plenary conferences, and thematic workshops. The biannual conferences organized by SOMA have also encouraged the exchange of ideas and experiences among professionals and students from different fields, locations, and institutions in Mexico. This is one of the main strengths of our Society, as the participants experience the broad scope of astrobiology, identify opportunities that can help them to initiate or consolidate a career in the field, and spend fruitful time with academic leaders from Mexico and other countries. Figure 9.3 is a collage of some posters used to promote SOMA Meetings and Conferences.

In August 2021, SOMA organized and hosted the III Latin American Conference of Astrobiology (Tercer Congreso Latinoamericano de Astrobiología, 3CLA), which was a virtual event due to the health contingency caused by the SARS-CoV2 virus (COVID-19). The 3CLA was organized in honor of Rafael Navarro González, one of the forerunners of Astrobiology in Mexico.

9.3.1.2   Mexican School of Astrobiology

The event that is now identified as the Escuela Mexicana de Astrobiología (EMA) had its origins in the first workshop on astrobiology organized in Mexico City from August 22nd to November 29th in 2009 at Facultad de Ciencias, UNAM. The workshop consisted of talks and activities for students and professionals interested in the field and was very well welcomed. Two years later, SOMA organized the first EMA in June 2011. Since then, the Schools have followed a biannual periodicity and have been hosted by different universities in Mexico. The sixth EMA was organized at Centro de Investigaciones Químicas, UAEM, in Cuernavaca, Morelos, in September 2022. Figure 9.4 is a collage of posters used to promote some of the EMA's.

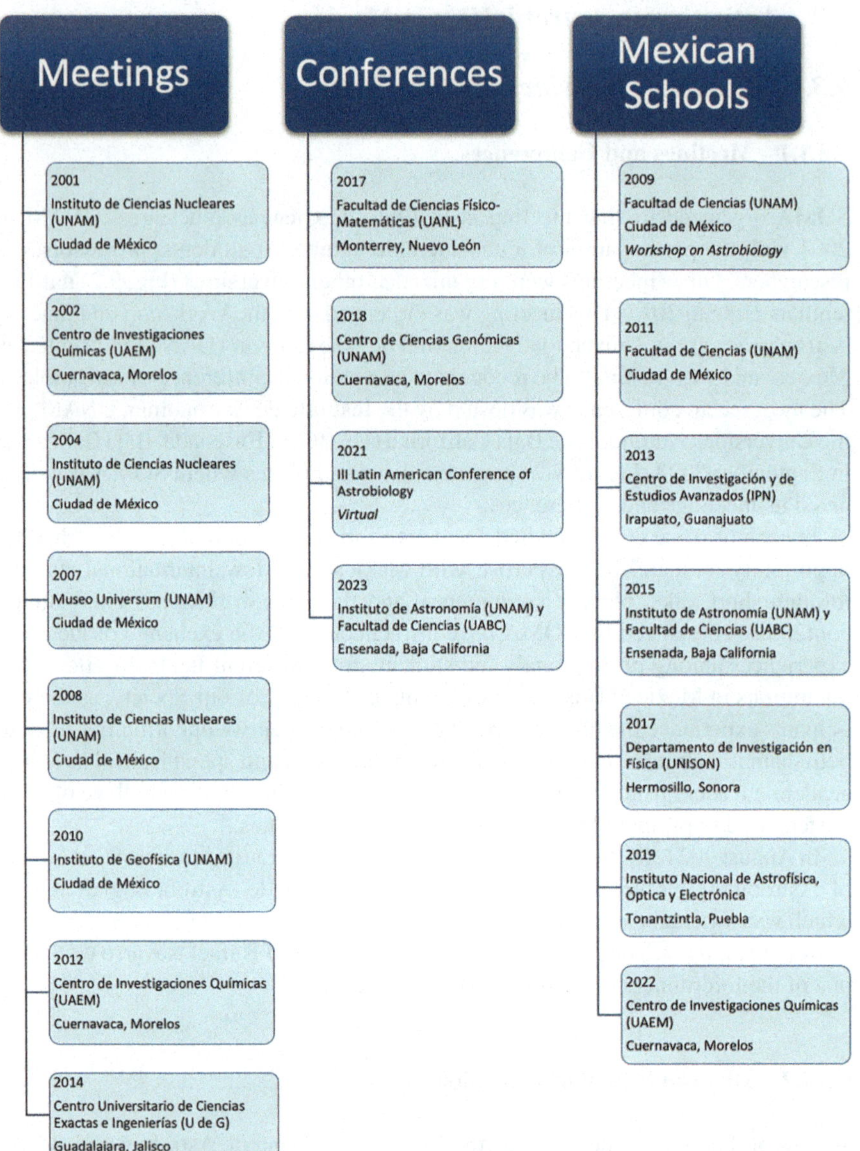

Fig. 9.2 Venues and year where SOMA has organized its main academic events

The Mexican School of Astrobiology is devoted to students. A national open call is published, and applications are received from people who are either in the last year of their undergraduate program (Licenciatura) or in the first year of their master's or Ph.D. program. Typically, a group of 30 students is selected based on their academic merits and motivation towards astrobiology. Thanks to the financial

Fig. 9.3   Posters used to announce some of the meetings and conferences organized by SOMA

support from institutions such as Consejo Nacional de Humanidades, Ciencias y Tecnologías (CONAHCYT, known before as CONACYT and currently as SECIHTI), Agencia Espacial Mexicana (AEM), Universities (UNAM, UAEM, UABC, UANL, UNISON), and the host institution, we are in the possibility to cover the travel, accommodation, and nurturing expenses of the selected students for 5 days. During this time, the students are entirely devoted to lectures, laboratory activities, and workshops on topics relevant to astrobiology. Also, they put into practice their knowledge and abilities on a field trip that occurs in a location of astrobiological interest, and most of the time, we conclude with an astronomical observation. All these activities are aimed to show the multidisciplinary character of astrobiology, and very quickly, it becomes a vivid experience that allows them to learn how to interact with students from very different academic backgrounds, how to perform collaborative tasks, and how they need to articulate all of this work to initiate an astrobiological research project.

Fig. 9.4 Posters used to announce some of the Escuela Mexicana de Astrobiología (EMA) organized by SOMA

9.3.1.3 Teaching

Undergraduate programs. All the academics associated with SOMA develop teaching activities on topics directly related to their primary professional training, such as chemistry, astronomy, or biology. However, they always emphasize the relationship of these topics with astrobiology.

The first astrobiology course was taught by a group of SOMA members in 2004 in the physics undergraduate program at Facultad de Ciencias of the Universidad Autónoma de Baja California (UABC). It was an introductory course that grew and evolved. More than 650 students have taken this course (Vázquez et al., 2019; Vázquez & Núñez, 2023).

At Facultad de Ciencias, UNAM, another astrobiology course was taught from 2008 to 2011, also in the physics undergraduate program that was then changed to the Ciencias de la Tierra undergraduate program and became a mandatory course to the spatial sciences orientation.

Since 2010, the biology undergraduate program offered by Facultad de Biología of the Universidad Autónoma del Estado de Morelos (UAEM) has hosted a more interactive astrobiology course, where traditional classes are complemented with activities posted in a virtual classroom.

In all cases, some students graduated after developing an astrobiologically-oriented thesis project.

Master and PhD programs. Astrobiology courses are offered under optative or specialization courses in different Master's and Doctorate programs such as the astrophysics program at UNAM, planetary astrophysics program at UANL, and sciences program at UAEM. SOMA academic members created all these courses, and some students have graduated with a thesis project where an astrobiological question or problem has been investigated.

Another aspect of the teaching activities is the habilitation of college teachers and university professors in some astrobiology topics. Under this scope, two workshops and one course have been organized. These activities will help us to achieve a broader and better-supported development of astrobiology in Mexico.

9.3.1.4 Astrobiology Book Written in Spanish

The book "Astrobiología. Una visión transdisciplinaria de la vida en el universo" (Fig. 9.5) is the most ambitious project generated by the academic members of SOMA. This is an electronic book with 13 chapters (Table 9.1) and is available from the Fondo de Cultura Económica (FCE) and UNAM.[2] This book provides scientific information on topics like the origin and evolution of life; the formation and evolution of planetary systems; the characteristics that made Earth a habitable planet; extreme environments and extremophiles; the astrobiological relevance of the planet Mars and that of the satellites Titan, Enceladus, Europa, the Moon; the discovery of exoplanets; the search for life in the universe, to mention a few. The book is written comprehensively, so that scholars from college to undergraduate levels find it a source of scientifically accurate information that is easy to appreciate. If more technical information is needed, many references are provided. Besides, the electronic book is written in Spanish, so it helps to fill the void that young students often face when they look for validated information about these topics (Montoya Lorenzana et al., 2022).

---

[2] Fondo de Cultura Económica (FCE), ISBN: 9786071674500: https://elfondoenlinea.com/Detalle.aspx?ctit=9786071674500

Libros UNAM, ISBN: 9786073068529: http://www.libros.unam.mx/astrobiologia-una-vision-transdisciplinaria-de-la-vida-del-universo-9786073068529-ebook.html

Fig. 9.5 Cover of the book "Astrobiología. Una visión transdisciplinaria de la vida en el universo" written in Spanish by academic members of SOMA (Montoya Lorenzana et al., 2022)

Hopefully, this book will be an interesting initial step for those interested in astrobiology and those who actively contribute to its consolidation in Mexico and other Spanish-speaking countries.

Table 9.1 Content of the book entitled "Astrobiology: a transdisciplinary vision about Life in the Universe" (Montoya Lorenzana et al., 2022)

Number	Chapter title	Author(s)
1	Astrobiology and its development in Mexico	Sandra Ignacia Ramírez Jiménez
2	The Universe	Leticia Carigi Delgado
3	Planetary Systems	Antígona Segura Peralta
4	The Earth as a habitable planet	Sandra Ignacia Ramírez Jiménez and María Guadalupe Cordero Tercero
5	The fossil record of minerals	Elizabeth Chacón Baca, Esther M. Cruz Gámez and Augusto A. Rodríguez Díaz
6	Life on Earth	Irma Lozada-Chávez
7	The origin of life on Earth	Irma Lozada-Chávez and Peter F. Stadler
8	Biological evolution	Irma Lozada-Chávez, Lilia Montoya Lorenzana, María Guadalupe Cordero Tercero, Cruz Lozano Ramírez and Julio E. Valdivia Silva
9	Extremophiles and extreme environments	Lilia Montoya Lorenzana and Sandra Ignacia Ramírez Jiménez
10	Astrobiology on Mars	María Guadalupe Cordero Tercero and Julio E. Valdivia Silva
11	Astrobiology of some of the Solar System's satellites	Sandra Ignacia Ramírez Jiménez, Lilia Montoya Lorenzana, María Guadalupe Cordero Tercero and Elva Escobar Briones
12	Life in the Universe	Antígona Segura and Leticia Carigi Delgado
13	The astrobiological meaning of life on Earth	Irma Lozada-Chávez and Roberto Aretxaga-Burgos

9.3.2 Outreach Activities

9.3.2.1 Talks

All researchers, members of SOMA, participate in giving talks about our lines of work related to Astrobiology. These talks have been presented at science fairs, elementary schools, high schools, and universities covering a wide range of audiences. In 2012, a series of talks was organized for young men at a juvenile detention center. SOMA members are convinced that it is very important to share science and daily work not only with students of different educational levels but also with the general public, because providing accurate information helps to destroy myths, reduce fears, avoid falling into charlatanism, and make better decisions. In addition, we encourage the interest of everybody in science, in particular, we want to motivate girls and young women to pursue scientific careers. To tell people what we do, it is also important to inform them why and how we research and its relevance for society.

Since 2021, SOMA has organized a series of talks called "Astrobiología para tod@s" (Astrobiology for everyone), as well as other talks, which can be watched— in some cases with translation into Mexican sign language—on its Facebook page and YouTube channel (Fig. 9.1).

9.3.2.2 Other Outreach Activities

Since 2009, the Mexican Society of Astrobiology has participated uninterruptedly in the event called "The Night of the Stars", organized by the Instituto de Astronomía of the Universidad Nacional Autónoma de México (UNAM). SOMA participates attending one of the several stands set up for this purpose either in the gardens of the University or in the main square of Mexico City. Also, this event is carried out in Ensenada, Baja California, for the colleagues who work in that City.

In addition to "The Night of the Stars", SOMA participates in "The Encounter with the Earth", an activity carried out in the Alameda of Santa María la Ribera, in Mexico City, organized by several institutes dedicated to Earth sciences. The objective of this event is to disseminate terrestrial geological phenomena such as earthquakes, volcanoes, groundwater, geomagnetism, ecology, etc.

Since June 2016, SOMA has joined the activities of the "Asteroid Day" (asteroidday.org) and some of its members are the local organizers in Mexico. The Society has participated in other Science fairs such as the Cordoba Science Fair, organized by the municipality of Cordoba, Veracruz, México, in 2015 and in the "Noche de las Ciencias" in Ensenada, Baja California.

SOMA's participation in the various science outreach events is supported by the following materials:

1. *Posters.* In each outreach event, we support our activities with a series of posters with topics such as Astrobiology, extremophiles and extreme environments, life in the Solar System, terraforming of Mars, exoplanets, habitable worlds, chemical evolution of the universe and life as we know it, and asteroids between others.

2. *Romboflexagons.* It is another type of flexagon (Fig. 9.6a), in which three different models have been developed with the themes "Space Missions", "Earthquakes in the Solar System", and "Astrobiology". The latter is shown in Fig. 9.6a.

3. *Hexahexaflexagons* (Fig. 9.6b). Based on the flexagon proposed by Arthur Stone in 1939. Images, designed exclusively for this purpose, were incorporated instead of numbers in the Stone's flexagons. The themes represented on its six faces are the SOMA's logo, exoplanets, extremophiles, Jupiter and Europa, Saturn and Titan and terraforming of Mars. It is very pleasing to see the interest and amazement caused by this flexagon in those who have played with it.

4. *Paper engineering cards.* There are two models of paper engineering cards, one in which a spiral of the solar system is formed and another that corresponds to

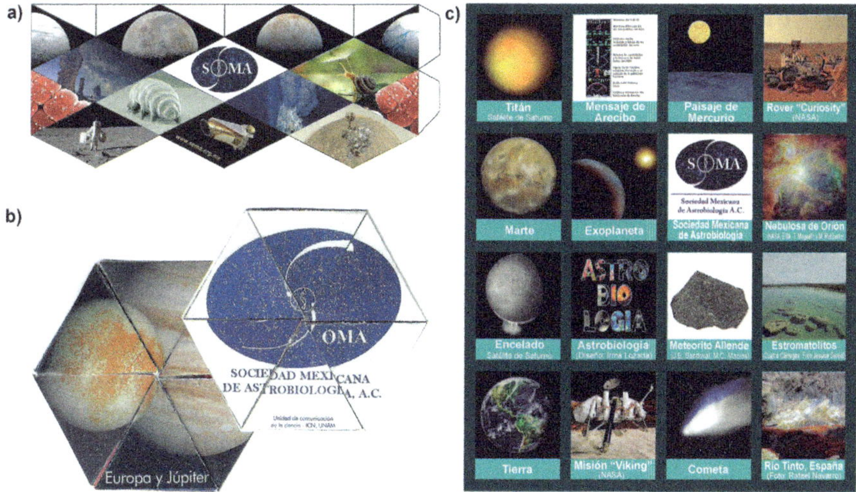

Fig. 9.6 Some of the materials used in outreach activities: (**a**) "Romboflexagon" with several astrobiological topics, (**b**) Hexahexaflexagon with six images representing: SOMA's logo, Europa and Jupiter, Terraformation, Exoplanet, Hyperthermophile, and Titan and Saturn, (**c**) One of the nine cards of the astrobiology lottery

"pop-up" cards with various themes (Earth, Moon, Sun, exoplanets, extremophiles, asteroids, and comets).

5. *Astrobiological Lottery.* It consists of nine boards and 54 cards (Fig. 9.6c), in which we show images related to Astrobiology that serve as a pretext to talk about different topics with the participants.

6. *Interactive games.* Among them are: "Put the exoplanet to the star" which is a variant of the game "Pin the tail on the donkey"; "Guess who" with images of scientists whose contributions have been relevant to Astrobiology; and two models of memory games, one of extremophile organisms and the other of images of galaxies, planets, satellites and minor planets.

7. *Manual demonstrations.* We have an activity called "Planets in a bottle" in which, using different recyclable materials, it is shown the probability of survival of yeast (used as an example of a terrestrial organism), when it is exposed to different environments similar to those observed in other planets and satellites of the solar system.

8. *Scale models.* There are four artistic models that show what the surface of four different exoplanets may look like. There is also a 60 cm long model of a tardigrade.

9. *Origami of flowers.* This activity explains what the color of plants would be like on some exoplanets that move around stars hotter or cooler than the Sun.

10. *Ask an Astrobiologist.* Members of SOMA answer questions from the general public all day long.

9.4   Field Sites of Astrobiological Interest in Mexico

Mexico ranks among the top five most megadiverse countries globally, hosting between 10 and 12% of the world's species and a significant proportion of endemic taxa (Mittermeier et al., 2005). The country also encompasses a wide array of diverse natural environments, including lakes, ponds, lagoons, crater lakes, basins, caves, volcanoes, geothermal fields, phreatic sinkholes (known as cenotes), and oceanic hydrothermal vents. Some of these extreme environments are considered analogs for the metabolic ecosystems of terrestrial life on early Earth, as well as on Mars, Europa (a moon of Jupiter), and Enceladus (a moon of Saturn). Additionally, several of these ecosystems may harbor extremophiles or preserve biosignatures that could be instrumental in detecting past or present life on other worlds (Segura et al., 2020; Valdivieso-Ojeda & Huerta-Díaz, 2016). Mexican and international astrobiologists, both individually and through collaborative efforts, have developed a diverse range of research interests in these field sites. Their investigations include exploring geochemical profiles, characterizing known taxa and members of the "rare biosphere", simulating prebiotic reactions experimentally, and testing extremophile tolerance in controlled laboratory settings. To enhance both national and international scientific collaborations, SOMA (the Mexican Society of Astrobiology) has dedicated significant resources to cataloging nearly 30 field sites of astrobiological interest throughout Mexico, alongside detailed information regarding the principal investigators conducting research at these locations. A comprehensive description of 12 selected field sites, based on the work of Cervantes de la Cruz et al. (2020), is presented below and illustrated in Fig. 9.7.

- The **Guaymas and Pescadero Basins**, in the Gulf of California, are hydrothermal systems featuring vent plumes, seeps, and anoxic sediments. Their microbial communities include bacterial mats, diverse prokaryotic and eukaryotic microbes, and symbiont-harboring invertebrates. These sites are used for testing underwater exploratory robots and as models for studying potential chemolitho-autotrophic communities in Europa, one of Jupiter's moons (Teske et al., 2016; Goffredi et al., 2017).
- **Figueroa Lagoon** (a.k.a. Laguna Mormona), in Ensenada, Baja California, is a closed and 5000-year-old hypersaline lagoon, almost completely filled with silica particles and gypsum. Laminated biofilms are located over large evaporite flats, which are found to be rich in oxygenic and anoxygenic phototrophs, as well as sulfate- and methane-reducing bacteria. Its conditions help redefine isotopic boundaries of biogenic methane, relevant for the detection of life on Mars (Franks & Stolz, 2009; Tazaz et al., 2013; Valdivieso-Ojeda & Huerta-Díaz, 2016).
- The **Salt Flats of Guerrero Negro**, in Baja California, consist of evaporative basins with high salinity (sodium and sulfate, 50–106 parts per thousand), where gypsum deposits and diverse microbial mat communities are formed. These environments contribute to our understanding of biomineralization processes, as well as of changes in the dominant methanogenic pathways and substrates used at extreme conditions (Vogel et al., 2009; Potter et al., 2009; Harris et al., 2013; Tazaz et al., 2013).

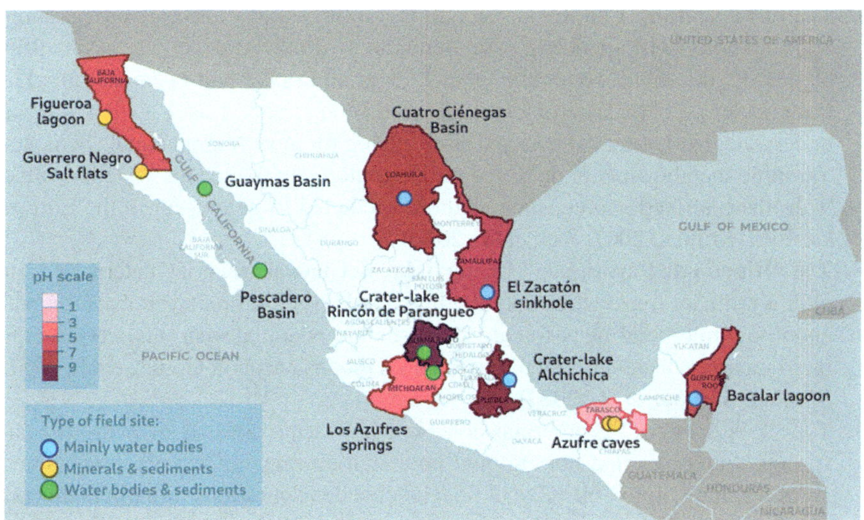

Fig. 9.7 Examples of field sites of astrobiological research interest in Mexico, with information from Cervantes de la Cruz et al. (2020) and references describing each site in the main text. Circles show the geographical location of 12 field sites across Mexico, which are colour coded according to the field type. States of the country hosting the field sites are colour coded according to the mean pH values measured on the sampling sites of astrobiological interests. The latitude and longitude coordinates for each site were retrieved from Google Maps (google.com/maps) on February 2025 and were plotted with Inkscape and the R-package *mxmaps* v2020.2.1 created by Valle-Jones (github.com/diegovalle/mxmaps): (1) Pescadero Basin, Gulf of California (pH: 6.5; 90–290 °C; 24.00001, −108.83334); (2) Guaymas Basin, Gulf of California (pH: 5.9–6.5; 25–350 °C; 27.24999, −111.49999); (3) Figueroa Lagoon (a.k.a. Laguna Mormona), Baja California (30.64497, −116.01896); (4) Salt flats of Guerrero Negro, Baja California (pH 6.9–7.9; 29–40 °C; 27.95409, −114.07644); (5) Pozas Azules, Cuatro Ciénegas Basin, Coahuila (pH: 8.5–8.8; −2.0–55 °C; 26.92259, −102.12245); (6) El Zacatón Cenote, Tamaulipas (pH: 6.5–6.7; 30 °C; 22.99292, −98.16560); (7) Rincón de Parangueo Crater-lake, Guanajuato (pH: 9.4–11.4; 19–26.8 °C; 20.43173, −101.25017); (8) Los Azufres geothermal filed, Michoacán (pH: 1–3 and 5.5–7.4; 27–87 °C; 19.78122, −100.65656); (9) Alchichica Crater-lake, Puebla (pH: 8.8–10; 14–21 °C; 19.41509, −97.40303); (10) Del Azufre Cave (a.k.a. Cueva de Villa Luz or Cueva de las Sardinas Ciegas), Tabasco (pH: 0–1.5; 25–33 °C; 17.44494, −92.76540); (11) Luna Azufre Cave (a.k.a. Sulfur Moon Cave), Tabasco (pH: 1.9; 29–30.3°C; 17.44494, −92.76540); (12) Bacalar Lagoon, Quintana Roo (pH: 7.7–8.2; 28–31 °C; 18.71042, −88.37189)

- The **Cuatro Ciénegas Basin**, in the Coahuila Desert, hosts over 300 highly diverse springs, streams and pools under a phosphorus and nitrogen-limited environment. It is one of the most microbially diverse sites on Earth, which is dominated by diverse microbialites with abundant ancient species, and a biological endemism similar to that of the Galapagos Islands. This site has been proposed as an analog for the Late Precambrian on Earth and the Gale Crater on Mars, including the identification of biosignatures of past and present biota (Souza et al., 2011, 2012; Centeno et al., 2012).

- The **"El Zacatón" Cenote**, in Tamaulipas, is the world's deepest vertical phreatic sinkhole (318 m deep), with remnant hydrothermal activity and sulfur clouds. Its microbial communities include phototrophic and chemoautotrophic bacteria, as well as anaerobic sulfide oxidizers. It serves as a testing site for underwater exploratory robots and as an analog for modeling potential chemoautotrophic metabolisms in deep and dark extraterrestrial aqueous systems with a hydrothermal feed source, such as those predicted to be present in the Europa satellite (Krajick, 2007; Sahl et al., 2010).
- The **"Rincón de Parangueo" Crater-lake**, in Guanajuato, is a Quaternary maar with a remnant hypersaline and alkaline pond. Hygroscopic cyanobacteria and endolithic microbial mats found here resemble potential microbial ecosystems on Mars and Enceladus, one of the Saturn's moons (Aranda-Gómez et al., 2017; Sánchez-Sánchez et al., 2023).
- The **"Los Azufres"** geothermal field, in Michoacán, consists of hydrothermal springs, fumaroles and boiling mud pools with a range of temperatures (some areas reaching up to 300 °C) and high arsenic concentrations. It harbors diverse mesophilic and extremophilic microorganisms, including sulfur oxidizing and reducing bacteria, aiding in the characterization of biosignatures for potential extraterrestrial life (Brito et al., 2014; Chen et al., 2018).
- **Alchichica Crater-lake**, in Puebla, is a soda, hyposaline and alkaline crater-lake with an oversaturation of magnesium and calcium carbonates. It hosts living microbialites dominated by cyanobacteria, anoxygenic phototrophs, and sulfate reducing bacteria. Its study is crucial for identifying biosignatures in Martian carbonates (Armienta et al., 2008; Navarro et al., 2010; Couradeau et al., 2011).
- **"Del Azufre"** and **"Luna Azufre" Caves**, in Tabasco, are permanently dark environments located at a Cretaceous limestone formation that sustain chemoautotrophic extremophiles. The "Luna Azufre" (a.k.a. Sulfur Moon Cave) contains ~500 m of passage and is located 320 m southeast from the larger "Del Azufre" cave (~2 km of passage), which is commonly known as Cueva de Villa Luz or Cueva de las Sardinas Ciegas. Most sampling sites at the "Del Azufre" cave are characterized by high hydrogen sulfide concentrations, whereas "Luna Azufre" is mostly a non-sulfidic cave. These caves serve as analogs for potential microbial ecosystems on other planets (Hose & Pisarowicz, 1999; Jones et al., 2016; Jones et al., 2023).
- Finally, **Bacalar Lagoon**, in Quintana Roo, is an oligosaline karstic coastal system hosting the largest and oldest occurrence of extant microbialites from the early Holocene, with high genetic diversity and organic carbon content. Its study contributes to the identification of biosignatures and potential microbial analogs for Mars and ancient life on Earth (Gischler et al., 2008; Yanez-Montalvo et al., 2020).

An enormous microbial diversity and global ecosystem patterns are being unveiled at these Mexican field sites (*e.g.*, Valdespino-Castillo et al., 2018; De Anda et al., 2018). Several sphylogenetic groups are endemic or environment specific, such as the calcifying Pleurocapsales cyanobacteria (Couradeau et al., 2011) and Candidatus Abyssubacteria (Sahl et al., 2010) that are exclusively found in either alkaline microbialite (Alchichica Lake) or the dark and often anoxic subsurface biosphere

(Cenote "El Zacatón"), respectively. Also, several field sites exhibit species turnover and phylogenetic diversity changes with strong environmental gradients in depth, humidity (wet *vs* dry), seasons (summer *vs* winter) and metabolisms (aerobic *vs* anaerobic). Other studies report the presence of a microbial core carrying out essential metabolisms across long spatial-temporal scales (Valdespino-Castillo et al., 2018), while some microbial communities seem to remain robust—embracing diversity and co-metabolic interactions—along intense and extended time spans of environmental perturbation (De Anda et al., 2018).

9.5 International Associations

The NASA Astrobiology Institute (NAI), established itself as a 'virtual institute' on May 19th 1998, consisting of research groups selected through a call for proposals with membership lasting 5 years. The purpose of this organization was to promote and lead research related to Astrobiology and included several international entities as associates and affiliates. In September 2012, SOMA submitted a proposal to NAI to become an affiliate partner. The proposal outlined the objectives of the Mexican Society of Astrobiology, listed academic and outreach activities carried out by Society members, individually and collectively up to that point, and demonstrated the richness of Astrobiological interest sites in Mexico, as well as the variety of research works related to the field. In January 2013, the Society received acceptance into NAI.[3] According to the acceptance letter, the partnership would initially focus on three areas: research of Astrobiological interest sites in Mexico (including virtual field trips), exchange of students and young researchers between both societies, and translation of NAI-created materials into Spanish for use in both Mexico and the United States. During this time, the SOMA president was invited to attend NAI leaders' meetings at AbSciCon conferences. Finally, NAI ceased to exist in January 2020,[4] and its coordination functions were taken over by the NASA Astrobiology Program.[5] Although some of its original objectives were retained, the membership of affiliated international entities is no longer active.

On the other hand, SOMA's inclusion in the European Astrobiology Network Association (EANA) was formalized in 2011. This association comprises researchers and European associations working on Astrobiology-related topics. At EANA conferences, the SOMA president has been invited to participate in meetings of the EANA Executive Council.

Declarations

Competing Interests The authors declare no competing interests.

---

[3] https://en.wikipedia.org/wiki/NASA_Astrobiology_Institute

[4] NASA Astrobiology Institute. "The NASA Astrobiology Institute Concludes Its 20-year Tenure". Retrieved 23-02-2024: https://web.archive.org/web/20200604195019/https://nai.nasa.gov/articles/2019/12/20/the-nasa-astrobiology-institute-to-end-its-20-year-tenure/

[5] NASA Astrobiology Program: https://astrobiology.nasa.gov/research/astrobiology-at-nasa/

References

Aranda-Gómez, J. J., Cerca, M., Rocha-Treviño, L., Carrera-Hernández, J. J., Levresse, G., Pacheco, J., Yutsis, V., Arzate-Flores, J. A., Chacón, E., & Beraldi-Campesi, H. (2017). Structural evidence of enhanced active subsidence at the bottom of a maar: Rincón de Parangueo, México. *Geological Society, London, Special Publications, 446*(1), 225. https://doi.org/10.1144/SP446.1

Armienta, M. A., Vilaclara, G., De la Cruz-Reyna, S., Ramos, S., Ceniceros, N., Cruz, O., Aguayo, A., & Arcega-Cabrera, F. (2008). Water chemistry of lakes related to active and inactive Mexican volcanoes. *Journal of Volcanology and Geothermal Research, 178*(2), 249–258. https://doi.org/10.1016/j.jvolgeores.2008.06.019

Brito, E. M., Villegas-Negrete, N., Sotelo-González, I. A., Caretta, C. A., Goñi-Urriza, M., Gassie, C., Hakil, F., Colin, Y., Duran, R., Gutiérrez-Corona, F., Piñón-Castillo, H. A., Cuevas-Rodríguez, G., Malm, O., Torres, J. P., Fahy, A., Reyna-López, G. E., & Guyoneaud, R. (2014). Microbial diversity in Los Azufres geothermal field (Michoacán, Mexico) and isolation of representative sulfate and sulfur reducers. *Extremophiles, 18*(2), 385–398. https://doi.org/10.1007/s00792-013-0624-7

Centeno, C. M., Legendre, P., Beltrán, Y., Alcántara-Hernández, R. J., Lidström, U. E., Ashby, M. N., & Falcon, L. I. (2012). Microbialite genetic diversity and composition relate to environmental variables. *FEMS Microbiology Ecology, 82*(3), 724–735. https://doi.org/10.1111/j.1574-6941.2012.01447.x

Cervantes de la Cruz, K., Cordero-Tercero, G., Gómez Maqueo Chew, Y., Lozada-Chávez, I., Montoya, L., Ramírez Jiménez, S. I., & Segura, A. (2020). Astrobiology and planetary sciences in Mexico. In *Astrobiology and Cuatro Ciénegas basin as an analog of early earth* (pp. 31–74). Springer. https://doi.org/10.1007/978-3-030-46087-7_2

Chen, L. X., Méndez-García, C., Dombrowski, N., Servín-Garcidueñas, L. E., Eloe-Fadrosh, E. A., Fang, B. Z., Luo, Z. H., Tan, S., Zhi, X. Y., Hua, Z. S., & Martinez-Romero, E. (2018). Metabolic versatility of small archaea Micrarchaeota and Parvarchaeota. *The ISME Journal, 12*(3), 756–775. https://doi.org/10.1038/s41396-017-0002-z

Couradeau, E., Benzerara, K., Moreira, D., Gerard, E., Kaźmierczak, J., Tavera, R., & López-García, P. (2011). Prokaryotic and eukaryotic community structure in field and cultured microbialites from the alkaline Lake Alchichica (Mexico). *PLoS One, 6*(12), e28767. https://doi.org/10.1371/journal.pone.0028767

De Anda, V., Zapata-Peñasco, I., Blaz, J., Poot-Hernández, A. C., Contreras-Moreira, B., Gonzalez-Laffitte, M., Gamez-Tamariz, N., Hernandez-Rosales, M., Eguiarte, L. E., & Souza, V. (2018). Understanding the mechanisms behind the response to environmental perturbation in microbial mats: A metagenomic-network based approach. *Frontiers in Microbiology, 9*, 2606. https://doi.org/10.3389/fmicb.2018.02606

Franks, J., & Stolz, J. F. (2009). Flat laminated microbial mat communities. *Earth-Science Reviews, 96*(3), 163–172. https://doi.org/10.1016/j.earscirev.2008.10.004

Gischler, E., Gibson, M. A., & Oschmann, W. (2008). Giant Holocene freshwater microbialites, Laguna Bacalar, Quintana Roo, Mexico. *Sedimentology, 55*, 1293–1309. https://doi.org/10.1111/j.1365-3091.2007.00946.x

Goffredi, S. K., Johnson, S., Tunnicliffe, V., Caress, D., Clague, D., Escobar, E., Lundsten, L., Paduan, J. B., Rouse, G., Salcedo, D. L., & Soto, L. A. (2017). Hydrothermal vent fields discovered in the southern Gulf of California clarify role of habitat in augmenting regional diversity. *Proceedings of the Royal Society B: Biological Sciences, 284*(1859), 20170817. https://doi.org/10.1098/rspb.2017.0817

Harris, J. K., Caporaso, J. G., Walker, J. J., Spear, J. R., Gold, N. J., Robertson, C. E., Hugenholtz, P., Goodrich, J., McDonald, D., Knights, D., & Marshall, P. (2013). Phylogenetic stratigraphy in the Guerrero Negro hypersaline microbial mat. *The ISME Journal, 7*(1), 50–60. https://doi.org/10.1038/ismej.2012.79

Hose, L., & Pisarowicz, J. A. (1999). Cueva de Villa luz, Tabasco, Mexico: Reconnaissance study of an active sulfur spring cave and ecosystem. *Journal of Caves and Karst Studies, 61*(1), 13–21.

Jones, D. S., Schaperdoth, I., & Macalady, J. L. (2016). Biogeography of sulfur-oxidizing Acidithiobacillus populations in extremely acidic cave biofilms. *The ISME Journal, 10*(12), 2879–2891. https://doi.org/10.1038/ismej.2016.74

Jones, D. S., Schaperdoth, I., Northup, D. E., Gómez-Cruz, R., & Macalady, J. L. (2023). Convergent community assembly among globally separated acidic cave biofilms. *Applied and Environmental Microbiology, 89*(1), 01575–01522. https://doi.org/10.1128/aem.01575-22

Krajick, K. (2007). Robot seeks new life and new funding in the abyss of Zacatón. *Science, 315*, 322–324. https://doi.org/10.1126/science.315.5810.322

Mittermeier, R. A., Gil, P. R., Hoffmann, M., Pilgrim, J., Brooks, T., Mittermeier, C. G., Lamoreux, J., & Da Fonseca, G. A. B. (2005). *Hotspots revisited: Earth's biologically richest and most endangered terrestrial ecoregions: Conservation International* (p. 315). Sierra Madre.

Montoya Lorenzana, L., Cordero Tercero, G., & Jiménez, R. (2022). *Astrobiología. Una visión transdisciplinaria de la vida en el Universo (Ediciones Científicas Universitarias)*. Fondo de Cultura Económica y Universidad Nacional Autónoma de México.

Navarro, K. F., Navarro-Gonzalez, R., Alcocer, J., Escobar, E., Morales, P., Cienfuegos, E., Coll, P., Lisa-France, C., Raulin, F. F., Stalport, F., & Cabane, M. (2010). *Chemical signatures of life in modern stromatolites from Lake Alchichica, Mexico. Applications for the search of life on Mars* (p. 13). 38th COSPAR Scientific Assembly.

Navarro-González, R., Molina, M. J., & Molina, L. T. (1998). Nitrogen fixation by volcanic lightning in the early earth. *Geophysical Research Letters, 25*, 3123–3126. https://doi.org/10.1029/98GL02423

Navarro-González, R., Vargas, E., de la Rosa, J., Raga, A. C., & McKay, C. P. (2010). Reanalysis of the Viking results suggests perchlorate and organics at midlatitudes on Mars. *Journal of Geophysical Research: Planets, 115*, E12. https://doi.org/10.1029/2010JE003599

Potter, E. G., Bebout, B. M., & Kelley, C. A. (2009). Isotopic composition of methane and inferred methanogenic substrates along a salinity gradient in a hypersaline microbial mat system. *Astrobiology, 9*, 383–390. https://doi.org/10.1089/ast.2008.0260

Sahl, J. W., Fairfield, N., Harris, J. K., Wettergreen, D., Stone, W. C., & Spear, J. R. (2010). Novel microbial diversity retrieved by autonomous robotic exploration of the world's deepest vertical phreatic sinkhole. *Astrobiology, 10*, 201–213. https://doi.org/10.1089/ast.2009.0378

Sánchez-Sánchez, J., Cerca, M., Moguel, B. B., Muñoz-Velasco, I., Alcántara-Hernández, R. J., Carreón-Freyre, D., & Levresse, G. (2023). Geomicrobiology of the Rincon de Parangueo maar crater: Exploring the link between an evolving extreme environment and its potential metabolic diversity. *Freshwater Biology, 68*, 1818–1838. https://doi.org/10.1111/fwb.14191

Segura, A., Ramírez Jiménez, S. I., & Lozada-Chávez, I. (2020). What is astrobiology? In *Astrobiology and Cuatro Ciénegas Basin as an analog of early earth* (pp. 1–30). Springer.

Souza, V., Siefert, J., Elser, J. T., & Eguiarte, L. E. (2011). The Cuatro Cienegas Basin in Coahuila, Mexico: An astrobiological precambrian park and mars analogue. *Analogue Sites for Mars Missions: MSL and Beyond, 1612*, 6007.

Souza, V., Siefert, J. L., Escalante, A. E., Elser, J. J., & Eguiarte, L. E. (2012). The Cuatro Ciénegas Basin in Coahuila, Mexico: An astrobiological Precambrian Park. *Astrobiology, 12*(7), 641–647. https://doi.org/10.1089/ast.2011.0675

Tazaz, A. M., Bebout, B. M., Kelley, C. A., Poole, J., & Chanton, J. P. (2013). Redefining the isotopic boundaries of biogenic methane: Methane from endoevaporites. *Icarus, 224*(2), 268–275. https://doi.org/10.1016/j.icarus.2012.06.008

Teske, A., de Beer, D., McKay, L. J., Tivey, M. K., Biddle, J. F., Hoer, D., Lloyd, K. G., Lever, M. A., Røy, H., Albert, D. B., Mendlovitz, H. P., & MacGregor, B. J. (2016). The Guaymas Basin Hiking Guide to Hydrothermal Mounds, Chimneys, and Microbial Mats: Complex Seafloor Expressions of Subsurface Hydrothermal Circulation. *Frontiers in Microbiology, 7*, 75. https://doi.org/10.3389/fmicb.2016.00075

Valdespino-Castillo, P. M., Hu, P., Merino-Ibarra, M., López-Gómez, L. M., Cerqueda-García, D., González-De Zayas, R., Pi-Puig, T., Lestayo, J. A., Holman, H. Y., & Falcón, L. I. (2018). Exploring biogeochemistry and microbial diversity of extant microbialites in Mexico and Cuba. *Frontiers in Microbiology, 9*, 510. https://doi.org/10.3389/fmicb.2018.00510

Valdivieso-Ojeda, J., & Huerta-Díaz, M. (2016). Microbial mats in Laguna Figueroa: Possible analogs of ancient life on Mars. *Mediterranews, 4*, 9–12.

Vázquez, R., & Núñez, P. G. (2023). 20 años de la enseñanza formal de la Astrobiología en Ensenada. *XII Congreso Nacional de Astrobiología, No., 3607*, 38.

Vázquez, R., Núñez, P. G., & Peña-Salinas, M. E. (2019). Experiences on teaching astrobiology in Baja California: From classical lectures to MOOCs. *Memorie della Societa Astronomica Italiana, 90*, 693–694.

Vogel, M. B., Des Marais, D. J., Turk, K. A., Parenteau, M. N., Jahnke, L. L., & Kubo, M. D. (2009). The role of biofilms in the sedimentology of actively forming gypsum deposits at Guerrero Negro. *Mexico. Astrobiology, 9*(9), 875–893. https://doi.org/10.1089/ast.2008.0325

Yanez-Montalvo, A., Gómez-Acata, S., Águila, B., Hernández-Arana, H., & Falcón, L. I. (2020). The microbiome of modern microbialites in Bacalar lagoon. *Mexico. PLoS ONE, 15*(3), e0230071. https://doi.org/10.1371/journal.pone.0230071

Chapter 10
Astrobiology in Peru: Early Achievements

Julio Valdivia-Silva

Abstract Astrobiology seeks to answer three fundamental questions: What are our origins and how have we evolved? Does life exist beyond Earth? And what is the future of life, both on our planet and elsewhere? Scientists aim to investigate these questions through seven key objectives that must incorporate various scientific disciplines to generate new knowledge. It is therefore important to highlight the contributions of Latin American countries to the scientific community in each of these areas. This section briefly describes the contributions of Peruvian astrobiologists—from the establishment of the Astrobiology Group of Peru (GAP) to the formation of the *Sociedad Científica de Astrobiología del Perú*, SCAP (Scientific Society of Astrobiology of Peru)—a journey marked by significant challenges and an inspiring example for young scientists in a country where science is not a top national priority. Nonetheless, astrobiology has enabled the unprecedented integration of multiple disciplines into joint research projects.

10.1   What Is Astrobiology?

When questions arise about our origins, the existence of life somewhere beyond our planet, or the future of life in the universe, it becomes difficult to rely solely on a textbook in biology, chemistry, physics, astronomy, or even geology to clearly answer them or to provide a comprehensive view of what we seek. Although each of these disciplines offers valuable concepts, none can independently provide a holistic and integrated answer to such fundamental inquiries. For this reason, astrobiology emerges as a new scientific field that integrates these diverse disciplines in a multidisciplinary effort to better address these essential questions from a scientific standpoint.

Julio Valdivia-Silva (✉)
Centro de Investigación en Bioingeniería, Universidad de Ingenieria y Tecnologia—UTEC, Lima, Peru
e-mail: jvaldivias@utec.edu.pe

D. Tovar, M. A. Leal (eds.), *Astrobiology and Planetary Sciences in Latin America*, https://doi.org/10.1007/978-3-032-01450-4_10

The multi-, inter-, and transdisciplinary nature of astrobiology enables a more comprehensive understanding of the phenomenon of life. Thanks to a roadmap outlined by NASA (Des Marais et al., 2008), the field now benefits from a structured framework for research and technological development, as well as a tool for systematic and academically grounded science communication.

While the goals and objectives established in that roadmap may take many years to be fully achieved, progress in each of them allows for the prioritization of research areas and the incorporation of new knowledge essential to each contributing discipline—thus continuing the exciting pursuit of our origins.

Finally, it is important to emphasize the fundamental principles that govern research in astrobiology (Des Marais et al., 2008):

1. Astrobiology is multidisciplinary in content and interdisciplinary in execution. Its success critically depends on close coordination among different disciplines and programs, including space missions.
2. Astrobiology promotes the stewardship of planetary processes, which involves protecting any potential biology that may be found on other planets while also safeguarding Earth by preventing contamination. This approach facilitates informed decision-making regarding ethical issues in planetary exploration.
3. Astrobiology acknowledges the broad societal interest in questions that may carry significant implications.
4. The public's intrinsic curiosity about these topics presents a crucial opportunity to educate and inspire new generations of scientists and informed citizens. Therefore, astrobiology serves as a powerful tool for science outreach and education.

10.2   Brief History of Astrobiology in Peru

During the decade from 2000 to 2010, the term "Astrobiology" in Peru was often confused with pseudoscientific fields by the general public and was scarcely disseminated within the scientific community. However, with the progress of Mars exploration missions such as *Spirit* and *Opportunity* (2004–2005), and *Phoenix* (2008), as well as studies demonstrating that Peru hosts a region considered an analog to the Martian surface (Valdivia-Silva et al., 2011), the concept began to gain popularity. In 2012, two students from the National University of San Marcos, Miguel Cervantes and Elizabeth Alanya, approached me with the idea of forming a group aimed at inspiring students through this emerging "multidisciplinary" field. It is important to note that astrobiology itself has been subject to ongoing debate over whether it constitutes a science or a multidisciplinary framework, as discussed in the previous section—though that remains a topic for another time. Both students were deeply motivated after attending the First Astrobiology Symposium held that year in the city of Arequipa, which had been organized by medical students from the National University of San Agustín for entirely science communication purposes.

Elizabeth and Miguel sought to expand this type of scientific event in Lima and to create an organization capable of supporting university students in training and eventually pursuing careers in the space sciences. This marked the founding of the organization "Astrobiology Group of Peru" (GAP).

The Astrobiology Group Peru (GAP) thus began its activities by focusing primarily on recruiting new members and promoting astrobiology in Peru through events featuring both national and international researchers. Activities included the organization of the Second Astrobiology Symposium of Peru and academic mentoring through videoconferencing. Between 2012 and 2015, university students from various institutions and myself as main advisor began to seek additional mentors specialized in astrobiology or related sciences to provide ongoing training and encourage the development of research projects. Initiatives such as International Astronomy Day and full sessions during the Encuentro Científico Internacional—an annual event that brings together Peruvian researchers from around the world—helped GAP gain popularity, drawing attention not only within Latin America but also internationally.

International scholars shared their experiences with local members through the Astrobiology Lectures, a program designed to promote knowledge exchange among members. In addition, various seminars, workshops, and space science classes in underserved rural schools played an important role in the popularization of the field. Starting in 2014, GAP began its internationalization process through external internships and participation in international events with support from The Mars Society International and the Mexican Astrobiology Society (SOMA), enabling Peruvian students to develop skills and acquire knowledge relevant to their professional formation in astrobiology.

GAP also launched expeditions to the Utah Desert (USA) and the Pampas de La Joya Desert (Peru)—both recognized as Martian analogs (Perez-Montaño & Valdivia-Silva, 2018; Fletcher et al., 2012; Valdivia-Silva et al., 2012). As a result, two new student organizations emerged in Peru that incorporated astrobiology into their missions: The Mars Society Peruvian Chapter (TMSP), with a primarily technological focus, and the Peruvian Association of Astrobiology (ASPAST) (Chon-Torres et al., 2020).

In 2015, GAP took a significant step forward by becoming formally established as the *Sociedad Científica de Astrobiología del Perú*, SCAP (www.scap.com.pe), founded upon three key pillars and with clearly defined goals focused on research, science communication, and education—both for its members and for the broader community. From that point on, I assumed the role of President of the first organization in Peru dedicated exclusively to the field of astrobiology.

Currently, SCAP continues its mission of outreach and education through workshops and seminars, as well as the organization of the First Latin American Congress of Astrobiology, a landmark event that fostered regional integration in this scientific field. The Society has reached significant milestones in research, as outlined below, and now has active chapters in five cities across Peru. All of these branches remain in close communication and work collaboratively toward the shared vision of establishing SCAP as a key hub for astrobiology in Latin America.

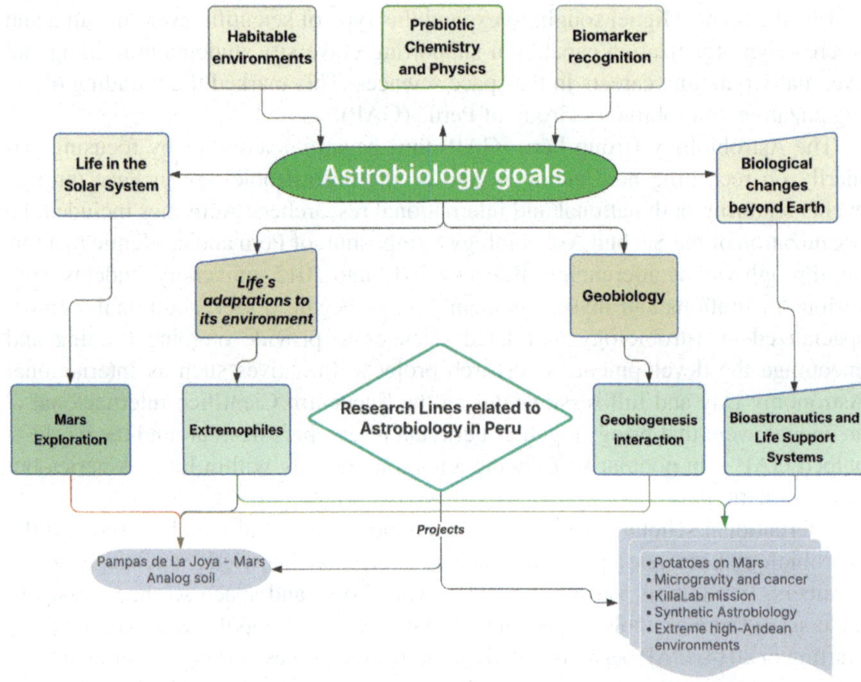

Fig. 10.1 Astrobiology goals and emerging research lines in Peru

10.3 Goals of Astrobiology and the Peruvian Contribution

The contributions in the most representative areas, based on the seven goals of Astrobiology, are summarized in Fig. 10.1, which highlights the projects that have achieved significant international recognition. Accordingly, in the following sections, we will provide a more detailed explanation of the projects referenced in that figure, while acknowledging the work of numerous researchers who have already made progress in various areas encompassed by these goals. Many of them are only now beginning to adopt an astrobiological approach by integrating and generating knowledge through multiple disciplines in response to the three fundamental questions posed in the first section.

10.3.1 Goal 1

10.3.1.1 Understanding the Nature and Distribution of Habitable Environments in the Universe

Determining the potential for habitable planets beyond our Solar System and characterizing those that are observable constitutes a central objective in the field of astrobiology. This goal opens the door to the search and exploration of Earth-like planets. We refer to our own planet because, to date, Earth remains the only known world that hosts life and is therefore considered habitable. Consequently, a habitable environment is defined as one that harbors living organisms or has the potential to do so. However, this definition presents a challenge: How can we determine whether a planet harbors life if we do not know what life may look like elsewhere? And how can we assess habitability in the absence of definitive evidence for the presence of living beings?

For this reason, the concept of habitability is currently framed around the conditions that would theoretically support or give rise to life in a given environment (Orgel, 1998a, 1998b). This definition is grounded in our biological experience and includes three principal criteria: (1) the presence of liquid water, which is the most effective solvent known for sustaining the chemical reactions underlying cellular metabolism. Several studies have demonstrated its superior properties compared to any other solvent (Baker et al., 2005; Brack & Pillinger, 1998); (2) the availability of conditions conducive to the assembly of complex carbon-based molecules, given that all biomolecules on Earth are composed of this element. Amino acids, sugars, nucleotides, and lipids are all carbon-based compounds. Although alternative elements have been theoretically proposed, thermodynamic analyses indicate that carbon remains the most favorable (Eiler et al., 1997); and (3) the existence of a viable energy source capable of driving chemical reactions and thereby sustaining cellular metabolism (Ponnamperuma et al., 1992; Southam et al., 2007).

Thus, when searching for habitable environments—such as planets orbiting a star—habitability is typically associated with a narrow orbital region known as the "stellar habitable zone." In this zone, the planet is located at a distance from its star that is neither too close (which would vaporize water) nor too far (which would cause it to freeze) (Bertaux et al., 2007; Peeters et al., 2009; Preston & Dartnell, 2014).

In recent years, the detection of extrasolar planets has increased dramatically, largely due to the Kepler Space Observatory, which operated outside Earth's atmosphere from March 2009 until 2018. Thanks to this telescope, over 2000 exoplanets were discovered during its operational lifetime.

In Peru, despite the presence of astronomical observatories, scientific research has not yet focused significantly on this particular field.

10.3.2 Goal 2

10.3.2.1 Identifying Past or Present Prebiotic Environments and Signs of Life in the Solar System

One of the central goals in astrobiology is to determine the existence of environments—past or present—that may have hosted prebiotic chemistry or preserved signs of life elsewhere in the Solar System. This includes reconstructing the history of environments that possessed liquid water, essential chemical ingredients, and energy sources capable of sustaining living systems. The search also involves analyzing surface and atmospheric materials that could contain evidence of past or extant life.

This objective aims to elucidate the distribution of the fundamental requirements for life throughout the Solar System and to understand how many potentially habitable environments may have persisted over geological time. Although life may have developed in forms different from those observed on Earth, the nature of life as we know it serves as a conceptual starting point. A compelling example is the discovery of extremophiles—organisms thriving in environments once thought uninhabitable. Life has been found in high-temperature zones, at great oceanic or crustal depths, under subzero conditions, in highly radioactive regions, in arid environments, and in acidic or alkaline ecosystems. These findings suggest that the so-called habitable zone in the Universe may be broader than previously assumed. Consequently, it is now considered plausible that the surface or subsurface of certain planets or moons in the Solar System may harbor hidden habitats, potentially occupied by microbial life forms, awaiting discovery with technologies first tested in terrestrial laboratories (Cavicchioli, 2002).

Among the most promising candidates for astrobiological exploration are Mars, Saturn's moons Titan and Enceladus, and Jupiter's moon Europa. Each of these celestial bodies exhibits unique features that suggest the possibility of supporting life, either on their surfaces or in subsurface environments.

Mars has long captivated scientific curiosity. As early as the 1900s, astronomers such as Giovanni Schiaparelli, in his work Life on Mars, illustrated the Martian surface with a series of lines he interpreted as canals, possibly built by a civilization similar to ours. Popular culture embraced this idea, leading to numerous stories and films about Martians seeking to invade Earth. However, it was not until the 1960s and 70s that NASA's Mariner orbiters captured detailed images of the Martian surface, revealing that Schiaparelli's canals were likely ancient riverbeds or fluvial channels (Coleman et al., 2007; Fassett & Head, 2008; Leighton et al., 1969). The Viking missions in 1975–1976 were the first to land on Mars and transmit images of its cold, desert-like surface. These missions provided critical data on Martian soil composition, mineralogy, and atmosphere, but also delivered disappointing results—no apparent organisms, no detectable water, and, most significantly, no trace of carbon. Attempts to detect photosynthetic or heterotrophic bacteria yielded inconclusive results. Consequently, Mars was initially deemed uninhabitable.

Nearly two decades later, the European Space Agency's Mars Express orbiter captured an extraordinary image indicating the presence of subsurface frozen water. This discovery reignited global interest in Mars. Subsequent missions—including Pathfinder, Spirit, Opportunity, Phoenix, and more recently, the Mars Science Laboratory (Curiosity)—provided compelling evidence that Mars may harbor multiple types of habitats within its interior (Chojnacki et al., 2014; Edwards & Ehlmann, 2015; Fan et al., 2008; McKay & Nedell, 1988; Wharton et al., 1995). Today, our understanding of Mars has evolved. Traces of carbon bonded with chlorine compounds have been detected (Ming et al., 2014), and evidence of liquid water—albeit highly saline due to perchlorates—has been observed (Martin-Torres et al., 2015). Both are essential indicators of potential habitability. Moreover, the availability of UV radiation as a possible energy source, the elevated temperatures in the subsurface, and the presence of salt and sulfate rocks—known on Earth to host microbial life—further strengthen the hypothesis of potential Martian microbial life (Davila et al., 2008; Wierzchos et al., 2006).

Saturn's moon Titan presents a distinctly different case. Due to its distance from Earth, only the Cassini–Huygens mission has yielded direct insights into its atmosphere. Titan is characterized by lakes of hydrocarbons, cryovolcanoes, streams of liquid nitrogen, and lightning—possible energy sources for prebiotic chemistry. The absence of oxygen prevents these lakes from igniting, but the mere presence of organic molecules raises the possibility of carbon-consuming microbial ecosystems. Titan may represent an analogue of early Earth, during the primordial stages of planetary evolution.

Another Saturnian moon, Enceladus, has also emerged as a compelling astrobiological target. Cassini–Huygens revealed geysers of water vapor erupting from its surface—evidence of subsurface water pockets maintained in liquid form by tidal forces induced by Saturn's gravity and orbital resonance. Even more significant is Jupiter's moon Europa, hypothesized to possess a vast subsurface ocean beneath approximately 10 kilometers of ice. Data from the Galileo mission indicate that this ice is geologically young, suggesting that it periodically fractures, allowing liquid water to reach the surface and renew the icy crust.

These celestial bodies represent only a fraction of the many intriguing candidates that astrobiology now examines through multidisciplinary approaches. Their study exemplifies how the convergence of biology, geology, chemistry, physics, and planetary science advances our understanding of potential life beyond Earth.

In Peru, this objective has been actively pursued through research conducted in the Pampas de La Joya, a terrestrial Mars analog site. Microbiological, geological, geomorphological, paleontological, pedological, climatological, chemical, and physical studies have produced a substantial body of scholarly publications and spurred new collaborative research initiatives, as outlined in Sect. 10.4 (Fig. 10.2).

10.3.3 Goal 3

10.3.3.1 Understanding How Life Emerges from Cosmic and Planetary Precursors

Theoretical, experimental, and observational approaches aimed at uncovering the chemical and physical principles underlying the origin(s) of life.

The origin of life remains an unresolved mystery that astrobiology approaches through multiple hypothetical pathways, with the terrestrial origin being just one among several possibilities. A foundational question in this field is identifying which materials could assemble into complex structures capable of undergoing chemical evolution on a planetary body or celestial object.

Properties evaluated on Mars		Hawaii	Salten Skov	Atacama (Yungay)	Mojave	Rio Tinto	El Jaroso	Dry Valleys	Atacama (Pampas de la Joya)
Chemical									
	Dielectric constant	-	-	-	-	-	-	-	-
	Redox potential	-	-	✓	-	-	-	-	✓
	pH	-	✓	✓	-	✓	✓	-	✓
	Electrical conductivity	-	-	✓	-	-	-	-	-
	Volatiles	✓	-	-	✓	-	-	-	-
	Mineralogy	✓	✓	✓	✓	✓	✓	-	✓
Mechanical									
	Cohesion forcé	✓	-	-	-	-	-	-	-
	Internal friction angle	✓	-	-	-	-	-	-	-
Physical									
	Particle size	✓	✓	-	-	-	-	✓	-
	Particle shape	-	-	-	✓	-	-	-	-
	Relative density	✓	-	-	-	-	-	✓	-
	Bulk density	✓	-	✓	-	-	-	-	✓
	Porosity	✓	-	-	-	-	-	-	-
	Water content	✓	-	✓	-	-	-	✓	-
Geological									
	Morphology	-	-	✓	-	-	-	✓	✓
	Geological processes	✓	-	✓	✓	-	✓	✓	✓
	Layer depth	-	✓	-	-	-	-	✓	-
Thermophysical									
	Albedo	-	-	-	-	-	-	-	-
	Thermal inertia	-	-	-	-	-	-	-	-
Magnetic									
	Magnetic susceptibility	-	-	-	-	-	-	-	-
	Magnetic saturation	-	✓	-	-	-	-	-	-
Organic									
	Total organic Carbon	-	-	✓	✓	✓	-	✓	✓
	Molecular abundance	✓	✓	✓	-	✓	✓	✓	✓
	Microorganism count	✓	✓	✓	-	✓	✓	✓	✓

Fig. 10.2 Pampas de La Joya Compared to Various Martian Analogues on Earth. The different locations highlight their relevance for space exploration due to their similarity to Martian soil

This line of inquiry focuses on how simple carbon-based molecules gradually increased in complexity, eventually acquiring the structural and functional properties necessary for self-replication and mutual coordination—traits that define the earliest biological systems. Such systems must also develop the capacity to absorb energy and nutrients from their environment, catalyze chemical reactions, and replicate their biomolecular components.

Astrobiology investigates all conceivable forms of "biological building blocks" that could have contributed to the construction of these systems. This includes the study of chemical fossils—not limited to cellular fossils—and the reconstruction of ancient biochemical diversity in the search for the most primitive common ancestor.

Importantly, this research agenda spans a wide spectrum: from identifying the sources—both exogenous and endogenous—of the molecular precursors that gave rise to life's foundational compounds, to unraveling the origin of the first cell (Kopp & Kirschvink, 2008; Schopf et al., 2007; Westall, 1999).

In Peru, a growing interest in modeling and experimental studies of primitive molecules and their genesis has emerged, primarily influenced by the academic lineage of Dr. Alicia Negrón and Dr. Antonio Lazcano from UNAM, Mexico. Currently, several research groups are initiating work in this domain, with the aspiration that Peruvian contributions to this field will become increasingly significant in the near future.

10.3.4   Goal 4

10.3.4.1   Understanding How Life on Earth and Its Planetary Environment Have Co-evolved Through Geological Time

Investigating the relationship between Earth and its biota through the integration of geosciences and biosciences, revealing how life evolved, responded to environmental changes, and exerted planetary-scale modifications.

The interplay between extraterrestrial processes that have shaped Earth's geological evolution and the internal planetary dynamics present since its formation is crucial for reconstructing the conditions and critical moments when life emerged and began to interact with its environment. Studying how life adapts to diverse conditions enhances our ability to define, detect, and interpret biosignatures left not only on Earth but potentially beyond it.

Thus, the study of Earth's atmosphere, geological history, and catastrophic events—such as the Late Heavy Bombardment approximately 3.9 billion years ago—is essential to understanding the early development of life. Subsequent mass extinction events and drastic geological transformations that affected early cells or species are also addressed within this research goal (Nisbet & Sleep, 2001; Sleep et al., 1989).

Paleontological studies that document the transition from unicellular to multicellular organisms and the emergence of the Eukarya domain are analyzed to understand the trajectory that led to the present biosphere. While a paleontologist typically focuses on the geological context of fossils, an astrobiological perspective extends this analysis to include atmospheric composition, geochemistry, geophysics, and the identification of molecular biomarkers that are resistant to degradation and may serve as potential biosignatures in analogous environments elsewhere (Schopf et al., 2007).

An illustrative case concerns the role of solar radiation in the origin or diversification of living organisms. Approximately 4 billion years ago, the Sun emitted significantly less energy than it does today. Some theories suggest that Earth may have been entirely frozen, necessitating biological adaptations to cryogenic conditions. However, alternative studies propose that the atmosphere may have contained higher concentrations of greenhouse gases, which would have trapped heat and allowed the persistence of liquid water—critical for chemical reactions associated with life (Shaw, 2008; Warren et al., 2002).

In Peru, interest in geobiology has emerged from isolated studies in extreme environments and paleontological findings along the Peruvian coast. Nevertheless, a more explicit astrobiological perspective is taking shape, driven primarily by research conducted in the Pampas de La Joya, and increasingly in high-altitude salt flats and paleo-lake systems where marine incursions once occurred. Once again, various research groups are recognizing the Pampas de La Joya as a strategic site for advancing investigations in this area.

10.3.5 Goal 5

10.3.5.1 Understanding the Evolutionary Mechanisms and Environmental Limits of Life

Determining the genetic, molecular, and biochemical mechanisms that govern and constrain the evolution, metabolic diversity, and adaptation of life.

Life on Earth has achieved extraordinary diversity as a result of the dynamic interplay between genetic information, the richness of metabolic pathways, and the environmental changes that have co-evolved throughout planetary history. While natural selection has long been recognized as a central mechanism in evolutionary biology, it is also evident that microbial and multicellular communities have actively contributed to global change at the planetary scale. One of the most illustrative examples is the release of oxygen into the atmosphere. Although the event known as the Great Oxidation Event (or "Oxygen Holocaust") is now understood to have resulted from more than just the proliferation of photosynthetic bacteria, it nonetheless catalyzed the emergence of large multicellular organisms and a subsequent explosion of biological diversity (Dorado et al., 2010). These new populations began to modify their surroundings, triggering ecological processes such as landscape transformation and alterations in environmental parameters like temperature and pH.

This intrinsic relationship between living organisms and their environment—a remarkable property observed at all scales on Earth—has led to the presence of life in ecosystems once thought uninhabitable. It is crucial to clarify that "extreme" is defined from a human perspective, referring to conditions under which we cannot naturally survive. However, for extremophilic organisms, these are standard environmental conditions. Although these life forms rely on the same basic carbon-based molecules as all terrestrial life, they have evolved sophisticated molecular adaptations that allow them to thrive in highly complex settings, such as acidic or alkaline habitats, regions of intense radiation, extreme heat or pressure, or arid environments (Albers et al., 2006; Bloechl et al., 1997; Burton & Norris, 2000; Cavicchioli, 2002; Hendry, 2006; Neilson et al., 2012).

From an astrobiological perspective, understanding these adaptive mechanisms—together with the evolutionary history of the environments in which they emerged—could expand the definition of habitability both on Earth and beyond. For instance, the discovery of microorganisms capable of surviving in arid soils or terrains geochemically similar to those on Mars has prompted space agencies to target analogous habitats for exploration (Brack & Pillinger, 1998; Pikuta et al., 2007; Valdivia-Silva et al., 2011).

Such studies provide a baseline for investigating life's potential on other worlds—specifically in hidden, extreme niches that, on Earth, have proven that life can colonize any environment once it emerges.

In Peru, several research projects are currently focused on analyzing the adaptive mechanisms of extremophilic organisms living in salt flats and desert ecosystems, further highlighting the relevance of the Pampas de La Joya. Additional efforts assess the tolerance limits of different organisms in these extreme settings or in simulated space conditions. For example, the projects Potatoes on Mars, Microgravity and Cancer, and KillaLab for the Moon (Quispe-Pilco et al., 2019a), as well as studies in high-Andean ecosystems considered as extreme environments (Quispe-Pilco et al., 2019b; Quispe-Choque et al., 2023), are actively contributing to this field. Furthermore, emerging initiatives in synthetic biology seek to engineer novel bio-modified organisms capable of resisting such conditions.

10.3.6 Goal 6

10.3.6.1 Understanding the Principles That Shape Life on Earth and Beyond

Assessing the mechanisms that drive changes in ecosystems to predict their future under environmental perturbations, and exploring the potential of life to survive and develop in environments beyond Earth.

Astrobiology has undertaken major challenges aimed at understanding the future of life on Earth. Environmental changes driven by human activity, along with abrupt events in specific regions of the planet, exert significant pressure on living organisms, often demanding adaptive responses at rates far faster than those achieved through evolutionary processes over millions of years. Therefore, diverse lines of research that seek to connect environmental change with the biochemical mechanisms and ecological contexts of organisms are critical to predicting biological processes at multiple levels.

Investigations into how organisms adapt, survive, and evolve under space conditions are also encompassed within this objective. Understanding the survival strategies of organisms in space helps assess their potential for transfer between celestial bodies and contributes to the development of planetary protection protocols—both to prevent forward contamination of other worlds and to safeguard Earth from possible biological back-contamination.

Moreover, radiation and microgravity are two ever-present variables in the space environment. Determining their effects on living systems at the genetic, molecular, and cellular levels is essential if humanity is to realistically consider the establishment of extraterrestrial colonies (Blaber et al., 2010; Fitts et al., 2001; Grimm et al., 2014; Johnson et al., 2005).

It is important to emphasize that research in this area has significantly contributed to technological and biomedical advances on Earth. A noteworthy example is the understanding of osteoporosis acquired through spaceflight studies, which has led to the development of new treatments for this disease (Fitts et al., 2001).

The two major challenges in space research are the development of a sustainable artificial habitat that enables long-term human permanence, and the capacity to survive under two constant conditions: radiation and microgravity. To address these challenges, we rely on experimental tools that aim to resolve key questions about human biology in space. Unfortunately, most studies to date have focused more on phenotypic outcomes than on underlying genotypic responses. However, a growing trend in the coming years is the pursuit of a deeper understanding of gene expression under spaceflight conditions.

In Peru, three prominent projects are aligned with this vision: (1) *Potatoes on Mars*, (2) *Microgravity and Cancer*, and (3) *KillaLab*. These initiatives have significantly raised the visibility of the national scientific community, producing noteworthy results and generating considerable enthusiasm throughout Latin America.

10.3.7   Goal 7

10.3.7.1   Searching for the Most Reliable Biosignatures to Identify the Presence of Life on Other Worlds and in Early Earth

Recognizing these signals in past and present samples, or detecting them remotely in planetary atmospheres and distant surfaces.

The search for characteristic traces of life involves definitions grounded in current biological processes. The records left behind by organisms and their interactions with the environment are the only empirical examples we have of life. However, the diversity of such interactions generates a wide variety of traces—known as bio-signatures—that can serve as a baseline in the search for life, both on early Earth, when the first cells emerged, and in potential habitats beyond our planet.

Astrobiology thus investigates a range of candidates that may serve as selective indicators of biological presence. These biosignatures include chemical compounds that played essential roles in cellular structure or function, as well as molecular or chemical features that could only have been produced in the context of an ecosystem. Furthermore, the remote detection of characteristic features in planetary atmospheres or surfaces is a key component of this objective. One example is the identification of atmospheric oxygen or methane on a planet or moon distant from Earth. Although these gases may also result from abiotic processes, high concentrations could suggest biological activity.

Another important example is the use of Earth as an extrasolar analog—observing Earth from space as if it were an exoplanet, to determine which biosignatures could indicate the presence of life. Candidate molecules for remote detection include photopigments, such as various types of chlorophyll found in plants.

This objective strongly encourages the development of novel technologies for the sensitive detection of such biomarkers.

In Peru, several research teams such as *KillaLab* and *Atoq*—many of which include engineers—have launched significant projects focused on the design, modeling, and implementation of systems capable of detecting biological signals and supporting operations in confined environments. One such initiative is a CubeSat-based bioreactor that simulates Martian conditions and allows for experiments on the growth of crops such as potatoes. Other systems simulate microgravity environments. In a more multidisciplinary approach, the development of an experimental base named *MARS-PJ*—inspired by the Mars Desert Research Station (MDRS) in Utah—represents an ambitious proposal for Peru. This initiative has brought together not only national governmental institutions and individuals but also regional and global partners, including NASA, ESA, and JAXA.

10.4 Projects with Global Impact in Recent Years

10.4.1 *Pampas de La Joya as a Martian Analog*

The Pampas de La Joya Desert, located in southern Peru and northern Chile, has proven to be one of the driest places on Earth, with some of the lowest levels of organic matter recorded—comparable to those measured on Mars (Fig. 10.3). Currently, numerous research groups are investigating the presence of microorganisms and potential subsurface habitats where life might persist. Given the astrobiological context of these studies, a comprehensive understanding of the geological, chemical, physical, and environmental setting is required to address such questions.

Fig. 10.3 Image of the bioreactor (CubeSat) used for potato cultivation experiments under extreme conditions. The image shows potato growth in highly saline soils inside the reactor

The current priority for SCAP in this iconic site—recognized as a birthplace of astrobiology in Peru—is the political and legal protection of this natural laboratory. In recent years, the main access route into the desert, via the Arequipa–Mollendo highway, has become increasingly threatened by illegal land trafficking and occupation, which severely hinders access for research and conservation. In response, SCAP has initiated coordinated actions with national institutions, private stakeholders, and scientific societies across the region to implement support strategies aimed at preventing the site's degradation.

The most important strategic initiative is the construction of an experimental base designed to simulate a station on the Red Planet. This base is envisioned as a scientific and technological hub in southern Peru, serving as an open research center for Latin American space science. With the support of the scientific community, this vision is moving toward realization.

10.4.2 The "Potatoes on Mars" Project

Launched in 2015, the *Potatoes on Mars* project aimed to develop crops capable of growing in extreme environments, including soils where nothing appears to thrive. For both NASA and the International Potato Center (CIP), this goal was highly relevant—not only due to the threat of global warming and desertification across vast regions of Earth, but also as a step toward establishing human colonies on distant planets and moons. In early 2016, the project officially began in collaboration with the University of Engineering and Technology (UTEC) in Lima.

In Phase 1, the potato was selected as the primary candidate crop due to its remarkable adaptability. Potatoes have colonized diverse ecosystems across the globe, helped humanity endure historical famines, and exist in over 4500 varieties. A "Martian soil" analog was identified in the Pampas de La Joya, and a bioreactor (CubeSat) was constructed to replicate extreme conditions similar to those on Mars (Fig. 10.3).

Initial experiments were conducted using selected potato varieties while the bioreactor was under construction. Among the 64 varieties tested, four were identified as not only capable of growing in the Martian analog soil but also of producing tubers. Phase 2 involved using the bioreactor to simulate both Martian soil and atmospheric conditions. The results were published, showing that potatoes could tolerate high salinity conditions without major impediments (Ramirez et al., 2017).

Future stages of the project aim to determine the physiological and genetic limits of this crop and explore avenues for its enhancement, either through direct genetic improvement or through the management of soil salinity. Ultimately, the goal is to prepare for the cultivation of potatoes on Mars, potentially during a primarily robotic mission, and to support the long-term vision of a more self-sufficient human settlement beyond Earth.

10.4.3 Microgravity and Cancer Project

One of the main reasons for the limited understanding of microgravity's effects on living organisms lies in the high cost associated with the design, development, and launch of experimental platforms into space. The technical requirements of life support systems—such as those needed for cell or animal cultures—pose significant challenges, as the equipment used for such purposes is often large and therefore expensive. Moreover, due to the lack of fully automated systems, astronauts are frequently required to manually manage the equipment.

As a result, the development of systems capable of simulating microgravity on Earth has become a key challenge—particularly when these must be integrated with incubators or other platforms that sustain living tissues or organisms.

In Peru, preliminary studies using the fruit fly *Drosophila melanogaster* revealed notable changes in tumor cells, but not in healthy cells, under both simulated and actual microgravity conditions aboard the International Space Station. These findings prompted the return of Peruvian researchers previously involved in these investigations and led to the development of a terrestrial microgravity simulator in Peru. Experiments were then initiated using breast and colon cancer cell lines.

According to communication with the authors, significant genetic-level findings have emerged, offering new insights into why spaceflight induces such cellular changes. More importantly, these discoveries open new possibilities for identifying molecular targets that may help halt the progression of cancer.

10.4.4   Synthetic Astrobiology

Several research groups in Peru have recently begun exploring synthetic biology as a tool within astrobiological research. Synthetic biology represents a reengineering of what was formerly known as genetic engineering, with the notable addition of "standardized processes" that were absent in earlier approaches.

Thanks to this advancement, new technologies now use standardized DNA sequences—known as "biobricks"—as modular units that can be assembled into complex circuits and systems that respond to specific stimuli. The result is the creation of minimal cells that can, for instance, produce a desired compound without undergoing replication, or highly sensitive biosensors with far greater specificity than any previously known system.

Astrobiology explores how these synthetic systems could support human survival in extreme environments by producing essential compounds, protecting against radiation or cold, enabling terraforming, or reconstructing ancestral DNA sequences in laboratory settings. These investigations may even allow for the construction of pseudo-cells that contain precursors to DNA, offering new ways to explore the possible origins of life.

Such experiments, which once required highly complex infrastructure, are now increasingly feasible thanks to the integration of modern engineering tools. The incorporation of bioinformatics and the global network of digital fabrication laboratories (FabLabs) have facilitated paradigm shifts that are transforming the scope of research in synthetic astrobiology.

10.5   Conclusions

A more integrative vision of the three fundamental questions of astrobiology can ultimately be assessed and analyzed through the lens of the seven overarching goals, each encompassing different aspects of the origin, presence, and future of life in the Universe. Although some researchers do not yet recognize astrobiology as a stand-alone scientific discipline, it is precisely within this field—and only here—that diverse perspectives can be meaningfully debated and integrated to reach more comprehensive answers.

In Peru, astrobiology deserves increased attention, as the country possesses a number of scientifically valuable locations with high astrobiological potential. These include Martian analog soils (such as those found in the Pampas de La Joya), landscapes hosting extremophilic organisms (such as salt flats, glaciers, and thermal springs), areas with fossil records and ancient rock formations, and—most importantly—a strong and growing community of researchers and students. These individuals are developing new lines of investigation, including studies on microgravity and the genetic modification of organisms to enhance their resistance to space environments.

We hope that in the coming years, the Scientific Society for Astrobiology of Peru will further consolidate its role as a key educational initiative and expand its reach by inspiring children and youth toward science. The formation of a Latin American Astrobiology Organization is essential to achieving more sustainable development in the region. We look forward to seeing this vision become a reality in the near future.

Acknowledgements We extend our sincere gratitude to the organizers of the Second Latin American Astrobiology Congress, held in Bogotá, Colombia, and to the entire scientific community that supports these events in pursuit of collaborative and rigorous scientific work.

Declarations

Competing Interests The authors declare no competing interests.

References

Albers, S. V., Konings, W. N., & Driessen, A. J. M. (2006). Membranes of thermophiles and other extremophiles. *Extremophiles, 35*, 161–171. https://doi.org/10.1016/S0580-9517(08)70010-2

Baker, V. R., Dohm, J. M., Fairen, A. G., Ferre, T. P. A., Ferris, J. C., Miyamoto, H., & Schulze-Makuch, D. (2005). Extraterrestrial hydrogeology. *Hydrogeology Journal, 13*, 51–68. https://doi.org/10.1007/s10040-004-0433-2

Bertaux, J. L., Carr, M., Des Marais, D. J., & Gaidos, E. (2007). Conversations on the habitability of worlds: The importance of volatiles. *Space Science Reviews, 129*, 123–165. https://doi.org/10.1007/s11214-007-9193-3

Blaber, E., Marçal, H., & Burns, B. P. (2010). Bioastronautics: The Influence of Microgravity on Astronaut Health. *Astrobiology, 10*, 463–473. https://doi.org/10.1089/ast.2009.0415

Bloechl, E., Rachel, R., Burggraf, S., Hafenbradl, D., Jannasch, H. W., & Stetter, K. O. (1997). Pyrolobus fumarii, gen. and sp. nov., represents a novel group of archaea, extending the upper temperature limit for life to 113 degrees C. *Extremophiles, 1*, 14–21. https://doi.org/10.1007/s007920050010

Brack, A., & Pillinger, C. T. (1998). Life on Mars: Chemical arguments and clues from Martian meteorites. *Extremophiles, 2*, 313–319. https://doi.org/10.1007/s007920050074

Burton, N. P., & Norris, P. R. (2000). Microbiology of acidic, geothermal springs of Montserrat: Environmental rDNA analysis. *Extremophiles, 4*, 315–320. https://doi.org/10.1007/s007920070019

Cavicchioli, R. (2002). Extremophiles and the search for extraterrestrial life. *Astrobiology, 2*(3), 281–292. https://doi.org/10.1089/153110702762027862

Chojnacki, M., Burr, D. M., & Moersch, J. E. (2014). Valles Marineris dune fields as compared with other martian populations: Diversity of dune compositions, morphologies, and thermophysical properties. *Icarus, 230*, 96–142. https://doi.org/10.1016/j.icarus.2013.08.018

Chon-Torres, O. A., Ramírez, J. C. R., Rengifo, F. H., Vessoni, R. A. C., Laura, I. S., Ríos-Ruiz, F. G. A., et al. (2020). Attitudes and perceptions towards the scientific search for extraterrestrial life among students of public and private universities in Peru. *International Journal of Astrobiology, 19*(5), 360–368. https://doi.org/10.1017/S1473550420000130

Coleman, N. M., Dinwiddie, C. L., & Baker, V. R. (2007). Evidence that floodwaters filled and overflowed Capri Chasma, Mars. *Geophysical Research Letters, 34*, L07201. https://doi.org/10.1029/2006GL028872

Davila, A. F., Gomez-Silva, B., de los Rios, A., Ascaso, C., Olivares, H., McKay, C. P., & Wierzchos, J. (2008). Facilitation of endolithic microbial survival in the hyperarid core of the Atacama Desert by mineral deliquescence. *Journal of Geophysical Research-Biogeosciences, 113*, G01028. https://doi.org/10.1029/2007JG000561

Des Marais, D. J., Nuth, J. A., III, Allamandola, L. J., Boss, A. P., Farmer, J. D., Hoehler, T. M., Jakosky, B. M., Meadows, V. S., Pohorille, A., Runnegar, B., et al. (2008). The NASA astrobiology roadmap. *Astrobiology, 8*, 715–730.

Dorado, G., Hernández Molina, P., Morales, A., Rey Fraile, I., & Vásquez, V. F. (2010). Biological mass extinctions on planet Earth.

Edwards, C. S., & Ehlmann, B. L. (2015). Carbon sequestration on Mars. *Geology, 43*(10), 863–866. https://doi.org/10.1130/G36983.1

Eiler, J. M., Mojzsis, S. J., & Arrhenius, G. (1997). Carbon isotope evidence for early life. *Nature, 386*, 665–665. https://doi.org/10.1038/386665a0

Fan, C. J., Schulze-Makuch, D., Fairen, A. G., & Wolff, J. A. (2008). A new hypothesis for the origin and redistribution of sulfates in the equatorial region of western Mars. *Geophysical Research Letters, 35*. https://doi.org/10.1029/2007GL033079

Fassett, C. I., & Head, J. W. (2008). The timing of martian valley network activity: Constraints from buffered crater counting. *Icarus, 195*, 61–89. https://doi.org/10.1016/j.icarus.2007.12.009

Fitts, R. H., Riley, D. R., & Widrick, J. J. (2001). Functional and structural adaptations of skeletal muscle to microgravity. *Journal of Experimental Biology, 204*, 3201–3208. https://doi.org/10.1242/jeb.204.18.3201

Fletcher, L. E., Valdivia-Silva, J. E., Perez-Montaño, S., Condori-Apaza, R. M., Conley, C. A., & McKay, C. P. (2012). Variability of organic material in surface horizons of the hyper-arid Mars-like soils of the Atacama Desert. *Advances in Space Research, 49*(2), 271–279. https://doi.org/10.1016/j.asr.2011.10.001

Grimm, D., Wehland, M., Pietsch, J., Aleshcheva, G., Wise, P., van Loon, J., Ulbrich, C., Magnusson, N., Infanger, M., & Bauer, J. (2014). Growing Tissues in Real and Simulated Microgravity. *New Methods for Tissue Engineering Tissue Engineering Part B: Reviews.* https://doi.org/10.1089/ten.teb.2013.0704

Hendry, P. (2006). Extremophiles: There's more to life. *Environmental Chemistry, 3*, 75–76. https://doi.org/10.1071/ENv3n2_ES

Johnson, A. S., Golightly, M. J., Weyland, M. D., Lin, T., & Zapp, E. N. (2005). *Minimizing space radiation exposure during extra-vehicular activity* (pp. 2524–2529). Space Weather. https://doi.org/10.1016/j.asr.2004.05.008

Kopp, R. E., & Kirschvink, J. L. (2008). The identification and biogeochemical interpretation of fossil magnetotactic bacteria. *Earth-Science Reviews, 86*, 42–61. https://doi.org/10.1016/j.earscirev.2007.08.001

Leighton, R. B., Horowitz, N. H., Murray, B. C., Sharp, R. P., Herriman, A. G., Young, A. T., Smith, B. A., Davies, M. E., & Leovy, C. B. (1969). Mariner 6 television pictures: First report. *Science, 165*, 685–690. https://doi.org/10.1126/science.165.3894.685

Martin-Torres, F. J., Zorzano, M.-P., Valentin-Serrano, P., Harri, A.-M., Genzer, M., Kemppinen, O., Rivera-Valentin, E. G., Jun, I., Wray, J., Bo Madsen, M., et al. (2015). Transient liquid water and water activity at Gale crater on Mars. *Nature Geoscience, 8*, 357–361. https://doi.org/10.1038/ngeo2412

McKay, C. P., & Nedell, S. S. (1988). Are there carbonate deposits in the Valles Marineris, Mars. *Icarus, 73*, 142–148. https://doi.org/10.1016/0019-1035(88)90088-7

Ming, D. W., Archer, P. D., Glavin, D. P., Eigenbrode, J. L., Franz, H. B., Sutter, B., Brunner, A. E., Stern, J. C., Freissinet, C., McAdam, A. C., et al. (2014). Volatile and organic compositions of sedimentary rocks in Yellowknife Bay, Gale Crater, Mars. *Science, 343*, 1245267. https://doi.org/10.1126/science.1245267

Neilson, J., Quade, J., Ortiz, M., Nelson, W., Legatzki, A., Tian, F., LaComb, M., Betancourt, J., Wing, R., Soderlund, C., et al. (2012). Life at the hyperarid margin: Novel bacterial diversity in arid soils of the Atacama Desert, Chile. *Extremophiles, 16*, 553–566. https://doi.org/10.1007/s00792-012-0454-z

Nisbet, E. G., & Sleep, N. H. (2001). The habitat and nature of early life. *Nature, 409,* 1083–1091. https://doi.org/10.1038/35059210

Orgel, L. E. (1998a). The origin of life—A review of facts and speculations. *Trends in Biochemical Sciences, 23,* 491–495.

Orgel, L. E. (1998b). The origin of life—How long did it take? *Origins of Life and Evolution of the Biosphere, 28,* 91–96. https://doi.org/10.1023/A:1006561308498

Peeters, Z., Quinn, R., Martins, Z., Sephton, M. A., Becker, L., van Loosdrecht, M. C. M., Brucato, J., Grunthaner, F., & Ehrenfreund, P. (2009). Habitability on planetary surfaces: Interdisciplinary preparation phase for future Mars missions. *International Journal of Astrobiology, 8,* 301–315. https://doi.org/10.1017/S1473550409990140

Perez-Montaño, H., & Valdivia-Silva, J. (2018). Martian Analogue of Pampas de La Joya: An update and future implications. https://doi.org/10.3390/IECG_2018-05371.

Pikuta, E. V., Hoover, R. B., & Tang, J. (2007). Microbial extremophiles at the limits of life. *Critical Reviews in Microbiology, 33,* 183–209. https://doi.org/10.1080/10408410701451948

Ponnamperuma, C., Honda, Y., & Navarro-Gonzalez, R. (1992). Chemical studies on the existence of extraterrestrial life. *Journal of the British Interplanetary Society, 45,* 241–249.

Preston, L. J., & Dartnell, L. R. (2014). Planetary habitability: Lessons learned from terrestrial analogues. *International Journal of Astrobiology, 13,* 81–98. https://doi.org/10.1017/S1473550413000396

Quispe-Choque, K. G., Condori Mamani, E., Quispe Llano, A. L., & Peréz Montaño, H. S. (2023). Assessment of extreme environments at different altitudinal gradients in arequipa with astrobiological and biotechnological potential. In *International conference on power electronics and instrumentation engineering* (pp. 1–15). Springer Nature Switzerland. https://doi.org/10.1007/978-3-031-61956-4_1

Quispe-Pilco, R. E., Venturo, S. C. R., Cruz-Simbrón, R. L., Ramírez-Gramber, J. J., Vásquez-Ortiz, V. E., Leonardo Julián, C., Valdivia-Silva, J., & Pérez-Montaño, H. S. (2019a). Conformation of an astrobiology interdisciplinary research group: The "team killalab" case study. In *Proceedings* (Vol. 24, p. 2). MDPI. https://doi.org/10.3390/IECG2019-06197

Quispe-Pilco, R. E., Rodriguez, S. C., Cruz, R., Cortes, J. P. E., Iquira, J. F., Gutarra, A., Valdivia-Silva, J., & Montoya, H. (2019b). Endurance of colonies of Nostoc sp. from high Andean ecosystems to simulated UV radiation on extreme conditions. In *AGU fall meeting 2019.* AGU.

Ramirez, D., Kreuze, J., Amoros, W., Valdivia-Silva, J., et al. (2017). Extreme salinity as a challenge to grow potatoes under Mars-like soil conditions: Targeting promising genotypes. *International Journal of Astrobiology,* 1–7. https://doi.org/10.1017/S1473550417000453

Schopf, J. W., Kudryavtsev, A. B., Czaja, A. D., & Tripathi, A. B. (2007). Evidence of archean life: Stromatolites and microfossils. *Precambrian Research, 158,* 141–155. https://doi.org/10.1016/j.precamres.2007.04.009

Shaw, G. H. (2008). Earth's atmosphere—Hadean to early Proterozoic. *Geochemistry, 68,* 235–264. https://doi.org/10.1016/j.chemer.2008.05.001

Sleep, N. H., Zahnle, K. J., Kasting, J. F., & Morowitz, H. J. (1989). Annhilation of ecosystems by large asteroid impacts on the early Earth. *Nature, 342,* 139–142. https://doi.org/10.1038/342139a0

Southam, G., Rothschild, L. J., & Westall, F. (2007). The geology and habitability of terrestrial planets: Fundamental requirements for life. *Geology and Habitability of Terrestrial Planets,* 7–34. https://doi.org/10.1007/978-0-387-74288-5_2

Valdivia-Silva, J. E., Navarro-González, R., Ortega-Gutierrez, F., Fletcher, L. E., Perez-Montaño, H. S., Condori-Apaza, R., & McKay, C. P. (2011). Multidisciplinary approach of the hyperarid desert of Pampas de La Joya in southern Peru as a new Mars-like soil analogue. *Geochimica et Cosmochimica Acta, 75,* 1975–1991. https://doi.org/10.1016/j.gca.2011.01.017

Valdivia-Silva, J. E., Navarro-González, R., Fletcher, L., Perez-Montaño, S., Condori-Apaza, R., & Mckay, C. P. (2012). Soil carbon distribution and site characteristics in hyper-arid soils of the Atacama Desert: A site with Mars-like soils. *Advances in Space Research, 50*(1), 108–122. https://doi.org/10.1016/j.asr.2012.03.003

Warren, S. G., Brandt, R. E., Grenfell, T. C., & McKay, C. P. (2002). Snowball Earth: Ice thickness on the tropical ocean. *Journal of Geophysical Research-Oceans, 107*, 1–8. https://doi.org/10.1029/2001JC001123

Westall, F. (1999). The nature of fossil bacteria: A guide to the search for extraterrestrial life. *Journal of Geophysical Research-Planets, 104*, 16437–16451. https://doi.org/10.1029/1998JE900051

Wharton, R. A., Crosby, J. M., McKay, C. P., & Rice, J. W. (1995). Paleolakes on Mars. *Journal of Paleolimnology, 13*, 267–283. https://doi.org/10.1007/BF00682769

Wierzchos, J., Ascaso, C., & McKay, C. P. (2006). Endolithic cyanobacteria in halite rocks from the hyperarid core of the Atacama Desert. *Astrobiology, 6*, 415–422. https://doi.org/10.1089/ast.2006.6.415

Chapter 11
Latin American Astrobiology Network

María Angélica Leal, David Tovar, Ximena Abrevaya, Jimena Sánchez, and Julio Valdivia-Silva

M. A. Leal (✉)
Grupo de Ciencias Planetarias y Astrobiología GCPA, Universidad Nacional de Colombia &
Corporación Científica Laguna, Bogotá, Colombia

Biology Department, Faculty of Sciences, Universidad Nacional de Colombia,
Bogotá, Colombia

Geosciences Department, Faculty of Sciences, Universidad Nacional de Colombia,
Bogotá, Colombia
e-mail: maria.leal@corpolaguna.org

D. Tovar
Grupo de Ciencias Planetarias y Astrobiología GCPA, Universidad Nacional de Colombia &
Corporación Científica Laguna, Bogotá, Colombia

Geosciences Department, Faculty of Sciences, Universidad Nacional de Colombia,
Bogotá, Colombia

Unit of Geology, Universidad de Alcalá, Alcalá de Henares, Spain
e-mail: dftovarr@unal.edu.co

X. Abrevaya
Instituto de Astronomía y Física del Espacio, UBA-CONICET,
Ciudad Autónoma de Buenos Aires, Argentina

Facultad de Ciencias Exactas y Naturales, Universidad de Buenos Aires,
Ciudad Autónoma de Buenos Aires, Argentina

Argentinian Research Unit in Astrobiology (Astrobio.ar),
Ciudad Autónoma de Buenos Aires, Argentina

J. Sánchez
Grupo de Ciencias Planetarias y Astrobiología GCPA, Universidad Nacional de Colombia &
Corporación Científica Laguna, Bogotá, Colombia

Biology Department, Faculty of Sciences, Universidad Nacional de Colombia,
Bogotá, Colombia

J. Valdivia-Silva
Universidad de Ingeniería y Tecnología—UTEC, Lima, Perú

D. Tovar, M. A. Leal (eds.), *Astrobiology and Planetary Sciences in Latin
America*, https://doi.org/10.1007/978-3-032-01450-4_11

Abstract This chapter presents the origins, evolution, and institutional structure of the Latin American Astrobiology Network (Red Latinoamericana de Astrobiología), a collaborative initiative aimed at strengthening scientific cooperation across the region in astrobiology and planetary sciences. The early efforts to establish a regional network date back to the late 1990s with the organization of Ibero-American postgraduate schools in astrobiology and the participation of leading researchers from Latin America and Europe. This and other initiatives laid the groundwork for the formal creation of a network in 2018 during the Second Latin American Congress of Astrobiology, held in Bogotá, Colombia. Since then, the Network has organized successive congresses, expanded its membership across the region, and formalized its statutes, objectives, and operational structure based on national nodes and associated members. The chapter describes the governance model, foundational principles, and long-term objectives of the Network, emphasizing its role in promoting interdisciplinary research, education, public engagement, and international collaboration. The Latin American Astrobiology Network stands as a dynamic and inclusive platform for advancing the study of life in the Universe from a regional perspective.

11.1 Early Network-Building Efforts in Latin America

Collaborative network-building efforts in Latin America began in 1997, during a meeting held at the Abdus Salam International Centre for Theoretical Physics (ICTP) in Trieste, where three scientists dedicated to the research in astrobiology Julián Chela-Flores, a Venezuelan physicist, Joan Oró a Spanish biochemist, and, Guillermo Lemarchand an Argentinian physicist, began planning the first Ibero-American Graduate School in Astrobiology. This school took place in November 1999 at the Institute for Advanced Studies of the Simón Bolívar University in Caracas, Venezuela (Lemarchand, 2010). The school featured over 20 faculty members, and Spanish was designated as the official language of instruction (Chela Flores et al., 2000).

During this inaugural school, the creation of an Ibero-American Astrobiology Network was proposed, along with the publication of a university-level textbook in Spanish (Lemarchand, 2010). These initiatives led to the publication of the book titled Astrobiology: Origins from the Big Bang to Civilization, which was ultimately released in English (Chela Flores et al., 2000). However, the proposed network was never formally established. Additionally, the organization of a second Ibero-American school—envisioned to take place in Oaxaca, Mexico in 2002—remained on the agenda but could not be realized (Lemarchand, 2010).

Subsequently, between 2006 and 2008, several workshops and seminars were held in Brazil, bringing together more than one hundred scientists from the region with active interest in astrobiology (Lemarchand, 2010). In 2009, the UNESCO Regional Office for Science in Latin America and the Caribbean supported the second Ibero-American Graduate School in Astrobiology, which was held at the Universidad de la República in Montevideo, Uruguay (Lemarchand & Tancredi,

2010). This initiative culminated in the publication of the Spanish-language book Astrobiología: del Big Bang a las civilizaciones (Astrobiology: From the Big Bang to Civilizations) (Lemarchand & Tancredi, 2010).

11.2   A Latin American Network Is Consolidated

The Ibero-American astrobiology schools played a pivotal role in inspiring many early-career researchers and fostering the formation of synergistic collaborations. It was within this context that, in 2015, conversations stablished between two Latin American astrobiologists, Dr. Ximena Abrevaya from Argentina and Dr. Julio Valdivia Silva from Perú, led to the idea to further explore the possibility of establishing a formal Latin American network for astrobiology and planetary sciences.

As a result of this initiative, the First Latin American Astrobiology Congress was organized and held in 2016 in the city of Lima, Peru (Fig. 11.1). This inaugural event brought together researchers from Argentina, Brazil, Colombia, Chile, Mexico, and Peru, with nearly one hundred participants in attendance. The program also included a field trip to the Pampas de La Joya, a recognized Mars analog site.

As part of the first congress, several proposals were submitted to host the second edition of the event, being Bogotá, Colombia ultimately selected as the venue. The Second Latin American Astrobiology Congress, organized by the *Grupo de Ciencias Planetarias y Astrobiología GCPA*, took place in 2018 (Fig. 11.2) without setbacks,

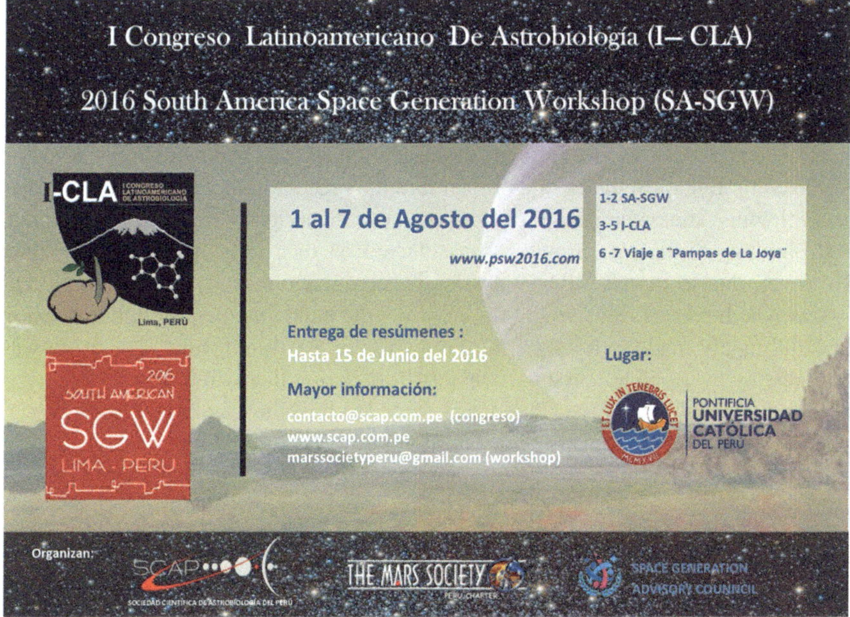

Fig. 11.1   First Latin American Astrobiology Congress, Lima, Peru, 2016

Fig. 11.2 Second Latin American Astrobiology Congress, Bogotá, Colombia, 2018

and brought together participants from Argentina, Brazil, Colombia, China, Spain, Mexico, Peru, and Uruguay—gathering approximately 150 attendees. The congress concluded a field trip to the Zipaquirá Salt Mine.

Among the topics discussed by the scientific community was the formal establishment and consolidation of the network that had been under development since 2015. As a result, statutes were drafted and the founding charter was signed on July 27, 2018, by Argentina, Colombia y Perú. There was also discussion regarding the possibility of producing a joint publication—a book that would document the history of astrobiology in Latin America. Work on this volume was initiated, and it was initially proposed that it be published by the *Universidad Nacional de Colombia*, the host institution of the Second Latin American Astrobiology Congress. However, due to administrative challenges stemming from the COVID-19 pandemic, the publication could not proceed through the university press. This situation led to the proposal to transform that initial project into the present book.

Following the submission of proposals to host the Third Latin American Astrobiology Congress, Mexico was selected as the venue for the year 2020. However, due to the COVID-19 pandemic, organizers initially opted to wait and assess whether conditions would improve. As the pandemic persisted, the decision was ultimately made to hold the third congress in a virtual format. The event was attended by over one hundred participants from across the region (Fig. 11.3).

For the Fourth Congress, Chile was selected to host the event in 2024. However, due to various circumstances, Chile later withdrew its commitment to serve as host. In light of the economic difficulties facing the region, it was decided to postpone the congress to 2025 or 2026. In parallel with these network-driven efforts, other countries in the region have also undertaken significant initiatives.

Fig. 11.3 Third Latin American Astrobiology Congress, Mexico City, Mexico, 2021

11.3 Objectives and Cooperation of the Latin American Astrobiology Network

Under the name LATIN AMERICAN ASTROBIOLOGY NETWORK, a civil network is established, governed by these statutes and the applicable laws and regulations. Its members shall be researchers affiliated with research groups, research centers, and universities from Latin American countries, with verifiable research activity in astrobiology, planetary sciences or in disciplines aligned with its goals and objectives.

General Objectives:
- Consolidate a recognized regional community for research, outreach, and education in astrobiology, planetary sciences and related sciences
- Provide a platform for regional and international collaboration, development, and advisory in astrobiology

Specific Objectives:
- Bring together researchers, groups, associations, institutions, and educational centers working on astrobiology-related topics
- Promote inter-institutional collaboration to strengthen research, outreach, and education efforts in line with the STEM principles (Science, Technology, Engineering, Mathematics) promoted by the United Nations.
- Support educational initiatives that foster youth interest in astrobiology and planetary sciences across the region.
- Promote ethical practices in astrobiology by overseeing the conduct of members and non-members alike
- Facilitate the creation of employment opportunities and professional environments for astrobiologists in Latin America.

To fulfill its objectives, the Network shall be empowered to:

- Strengthen collaboration among its members
- Seek and implement mechanisms in cooperation with agencies, private entities, or international partners to sustain long-term activities
- Encourage integration of academic and research organizations, and serve as a hub for cooperation and information exchange

- Establish mechanisms (in-person or virtual) to ensure the active participation of all stakeholders
- Organize and develop academic and scientific networks in Latin America
- Acquire, construct, or own movable and immovable property as needed
- Execute all necessary contracts, operations, and legal actions required to achieve its objectives
- Receive donations and funding from science and technology promotion agencies and international cooperation organizations
- Organize meetings, workshops, congresses, and other regional events to promote astrobiology activities and developments
- Maintain and publish up-to-date statistical information on the state of astrobiology in the region, aligned with the Network's areas of interest

11.4   Subsequent Affiliations and Current Members

Following the formal establishment of the Latin American Astrobiology Network in 2018, additional nodes and adherent members joined, especially from countries without a formally established node. The Network is currently composed as follows (see: www.astrobiologialatam.org):

- **Argentina Node:** *Núcleo Argentino de Investigación en Astrobiología (Argentinian Research Unit in Astrobiology)*
  Website: www.astrobioargentina.org
  Director: Ximena Abrevaya
- **Brazil Node:** *Sociedade Brasileira de Astrobiologia (Brazilian Society of Astrobiology)*
  Website: https://sbastrobio.org
  Current President: Gustavo Porto de Melo
- **Colombia Node:** *Grupo de Ciencias Planetarias y Astrobiología—GCPA*
  Website: www.corpolaguna.org/
  grupo-de-ciencias-planetarias-y-astrobiologia-gcpa
  Directors: Jimena Sánchez and David Tovar
- **Chile (Adherent Members):**
  Cecilia Demergasso—Universidad Católica del Norte
  Millarca Valenzuela—Universidad Católica del Norte
  Giovanni Leone—Universidad de Atacama
  Rómulo Oses—Universidad de Atacama
- **Cuba (Adherent Member):**
  Rolando Cárdenas—Universidad Marta Abreu de Las Villas
- **Mexico Node:** *Sociedad Mexicana de Astrobiología—SOMA*
  Website: https://sites.google.com/soma.org.mx/soma-a-c
  Current President: Sandra Ramírez

- **Peru Node:** *Sociedad Científica de Astrobiología del Perú—SCAP*
  Website: https://astrobiologyperu.wixsite.com/grupoastrobioperu
  Current President: Walter Guevara Day
- **Uruguay (Adherent Member):**

Gonzalo Tancredi—Universidad de la RepúblicaDeclarations

Competing Interests The authors declare no competing interests.

References

Chela Flores, J., Lemarchand, G. A., & Oró, J. (2000). *Astrobiology: Origins from the big bang to civilization*. Kluwer Academic Publishers.

Lemarchand, G. (2010). Una breve historia social de la astrobiología en Iberoamérica. In G. A. Lemarchand & G. Tancredi (Eds.), *Astrobiología: Del Big Bang a las Civilizaciones* (Vol. 1, pp. 23–52). Tópicos Especiales en Ciencias Básicas e Ingeniería.

Lemarchand, G., & Tancredi, G. (2010). Prefacio. In G. A. Lemarchand & G. Tancredi (Eds.), *Astrobiología: Del Big Bang a las Civilizaciones* (Vol. 1, pp. xi–xv). Tópicos Especiales en Ciencias Básicas e Ingeniería.

Made in the USA
Monee, IL
07 July 2026

56546136R00105